EUROPEAN EXTERNAL ACTION

Critical Geopolitics

Series Editors:

Alan Ingram, University College London, UK
Merje Kuus, University of British Columbia, Canada
Chih Yuan Woon, National University of Singapore, Singapore

Over the last two decades, critical geopolitics has become a prominent field in human geography. It has developed to encompass topics associated with popular culture, everyday life, architecture and urban form as well as the more familiar issues of security, inter-national relations and global power projection. Critical geopolitics takes inspiration from studies of governmentality and biopolitics, gender and sexuality, political economy and development, postcolonialism and the study of emotion and affect. Methodologically, it continues to employ discourse analysis and is engaging with ethnography and participatory research methods. This rich field continues to develop new ways of analysing geopolitics.

This series provides an opportunity for early career researchers as well as established scholars to publish theoretically informed monographs and edited volumes that engage with critical geopolitics and related areas such as international relations theory and security studies. With an emphasis on accessible writing, the books in the series will appeal to wider audiences including journalists, policy communities and civil society organizations with interests in international affairs and practices of security, identity and power.

Other books in this series

Europe in the World
EU Geopolitics and the Making of European Space
Edited by Luiza Bialasiewicz
ISBN 978 0 7546 7984 4

Reconstructing Conflict
Integrating War and Post-War Geographies
Edited by Scott Kirsch and Colin Flint
ISBN 978 1 4094 0470 5

Mapping the End Times
American Evangelical Geopolitics and Apocalyptic Visions
Edited by Jason Dittmer and Tristan Sturm
ISBN 978 0 7546 7601 0 / 978 1 4094 0083 7

European External Action

The Making of EU Diplomacy in Kenya

VEIT BACHMANN
Goethe-Universität Frankfurt am Main, Germany

Routledge
Taylor & Francis Group
LONDON AND NEW YORK

First published 2016 by Ashgate Publishing

Published 2016 by Routledge
2 Park Square, Milton Park, Abingdon, Oxon OX14 4RN
711 Third Avenue, New York, NY 10017, USA

First issued in paperback 2018

Routledge is an imprint of the Taylor & Francis Group, an informa business

British Library Cataloguing in Publication Data
A catalogue record for this book is available from the British Library

The Library of Congress has cataloged the printed edition as follows:
Names: Bachmann, Veit, 1979- author.
Title: European external action : the making of EU diplomacy in Kenya / by Veit Bachmann.
Description: Farnham, Surrey, UK ; Burlington, VT : Ashgate, 2016. | Series: Critical geopolitics | Includes bibliographical references and index.
Identifiers: LCCN 2015033458| ISBN 9781472423153 (hardback : alk. paper) |
Subjects: LCSH: European Union countries--Foreign relations--Kenya. | Kenya--Foreign relations--European Union countries. | European External Action Service.
Classification: LCC D2025.5.K46 B34 2016 | DDC 327.406762--dc23 LC record available at http://lccn.loc.gov/2015033458

ISBN 13: 978-1-138-54667-7 (pbk)
ISBN 13: 978-1-4724-2315-3 (hbk)

To Pommi

Contents

List of Figures, Tables and Textboxes

Figures

Table

Textboxes

Acknowledgements

I would like to express my gratitude to a number of people and institutions that have been instrumental for me in the process of developing this book. Primarily, I want to thank my partner Betty. 'Un grand merci à toi', Betty, for supporting me with your patience and love during the long fieldwork and grumpy writing phases and for enriching my thoughts and reflections on this book through your comments, ideas and your own work.

The German Ministry of Education and Research (BMBF) has generously provided funding for this research as part of the project *EuroGaps: External Relations and External Perceptions of the European Union in Eastern Europe and Sub-Saharan Africa*. And the Department of Human Geography at Goethe-University Frankfurt has functioned as an excellent host for our research group. I am highly indebted to all the informants and cooperation partners that have contributed to this research, in particular the Delegation of the European Union to Kenya and the University of Nairobi's Institute of Diplomacy and International Studies. Katy Crossan at Ashgate and Merje Kuus as series editor have provided invaluable support throughout the entire process of developing this book.

I am also very grateful to Alec Murphy, Anita Kiamba, Bernd, Christiane Tristl, Elke Alban, my brother Horsti and his family, Ievgenii Rovnyi, James Sidaway, Jason Dittmer, Johannes Fuchs, Makumi Mwagiru, Marc Boeckler, Martin Müller, Nannette Abrahams, my parents Günther and Christine, Öpfi, Peter Lindner, Sami Moisio, my sister Heike and her family, Stefan Ouma, Susan Amoko, Thomas Klinger and Till Paasche for supporting me in various ways during the course of this research project. All of your contributions have been of different nature, but helped me to complete this book through a mix of challenging and relaxing ways.

List of Abbreviations

AASM	Associated African States and Madagascar
ACP	African, Caribbean and Pacific Group
AMISOM	African Union Mission to Somalia
AU	African Union
BRICS	Brasil, Russia, India, China, South Africa
CFSP	Common Foreign and Security Policy
CSO	Civil Society Organisation
DG DevCo	Directorate General for Development Cooperation
EAC	East African Community
EBA	Everything-But-Arms
EC	European Community
ECC	European Economic Community
ECSC	European Coal and Steel Community
EDF	European Development Fund
EEAS	European External Action Service
EMU	European Monetary Union
EP	European Parliament
EPA	Economic Partnership Agreement
EPC	European Political Cooperation
EU	European Union
FRG	Federal Republic of Germany
FSI	Foreign Service Institute – Kenyan Foreign Ministry
GDR	German Democratic Republic
HoD	Heads of Delegation
HR/VP	High Representative of the Union Foreign Affairs and Security Policy/Vice President
ICC	International Criminal Court
IDIS	Institute of Diplomacy and International Studies – University of Nairobi
IO	International Organisation
JAES	Joint Africa-EU Strategy
LDC	Least Developed Country
NGO	Non-Governmental Organisation
NPE	Normative Power Europe
ODA	Official Development Assistance
PC	Political Counsellor
UK	United Kingdom

UNDP	United Nations Development Programme
US	United States
USA	United States of America
WTO	World Trade Organisation

Introduction
An Emerging EU Diplomacy?

The EU has long been described as an actor with substantial economic but limited political power (Twitchett 1976, 34). EU policy makers have also long attempted to increase the Union's political influence on a global stage. When the Treaty of Lisbon came into effect in 2009, it marked a key step in this process of the EU's political growth – both internally and externally. Part of a wider package of institutional reforms was the European External Action Service (EEAS), intended as a means for the EU to speak with one voice on a global stage (EEAS 2014a). With its external economic policy already deeply integrated, the EEAS focused on the political dimension of external relations, i.e. the integration of foreign policy. It sought to lay the foundation for developing a collective political role more attuned to the Union's economic strength. Reviewing her performance after four years of building the EEAS, the EU's foreign policy boss, Catherine Ashton, explained to the European Parliament in April 2014 that

> [w]e are unique in foreign policy terms. I don't just say this but you can hear this said across the world because we bring together economic, diplomatic, development and military assets in support of a comprehensive approach to complex crises. This approach did not exist 4 years ago and it exists now. And that increasingly makes us the partner of choice. (EEAS 2014d)

In the speech, Ashton points to the evolution in building the EU's 'assets', as she terms it, in multiple policy fields towards a 'comprehensive approach'. Ashton is convinced that this approach also makes the EU the 'partner of choice' based on what is being 'said across the world'. That Ashton's opinion about the 'new' international role for the EU, through the EEAS, might be more on the optimistic side of the spectrum of opinions on the EEAS is reflected in the nature of her position as the one charged with establishing the service. Outsiders are often more sceptical. In contrast to Ashton's assessment, key Kenyan cooperation partners of the EU described the EEAS as 'an internal EU thing [that] we don't even notice. It just does not matter' (Author interview, July 2012). In particular as regards the Union's aspiring political role, Ashton's assessment of the EU as the 'partner of choice' has not necessarily permeated the thought of those the EU is to partner with. For instance, a senior official in the Kenyan Ministry of Foreign Affairs argued that 'getting involved in political issues is out of tune' for the EU. It 'is not their business […] political aspects should be left to the ambassadors of US, UK, Germany and so on' (Author interview, March 2012). Clearly, differences between internal and external perceptions of an institution and its wider impact appear frequently. Yet for the EU as a polity that has been struggling for decades to increase its political influence on a global stage, the way it is seen by 'others', what is 'said about it' and if it can be the 'partner of choice' is of especial importance.

As part of these processes of building the EU's 'comprehensive approach', the EEAS transformed the European 'embassies' around the world from 'Delegations of the European Commission' to 'Delegations of the European Union', adding a political mandate of the European Council, i.e. the member states, to the existing mandate of the Commission. It thus added an institutional basis to long-established intergovernmental initiatives for aligning foreign policies (Allen, Rummel, and Wessels 1982; Lodge 1993) in an attempt to consolidate global EU *actorness* in a multipolar world (Howorth 2010). Such actorness, according to Bretherton and Vogler (2006), is comprised of the three elements of opportunity, presence and capability. Opportunity thereby denotes the external context; presence captures the ability of the EU to exert influence beyond its borders; and capability signifies the ability to make use of such opportunity and presence (ibid., 2, 24–34). In their approach, Bretherton and Vogler pay particular attention to processes of change and ask 'which internal and external factors have permitted, promoted or constrained the development of the EU's roles in global politics; how and to what extent is the EU perceived as an actor by its various "audiences"?' (ibid., 13).

The Lisbon Treaty and the EEAS were the tentative endpoints of an intensive process of enlargement and institutional reform during the 2000s – aimed at enhancing EU actorness and preparing the EU as a global actor. The narrative of successful regional integration, positioning the EU as a model to others in the world, has long been key to the EU's global actorness (Bachmann and Sidaway 2009b; Manners 2010). The outbreak of the financial crisis in 2008, however, has put this narrative under strain (Jones 2012; Murphy 2013). High levels of unemployment and social and economic precariousness became associated with the monetary union and the macroeconomic inertia of a polity comprised of different economies with diverse interests and capabilities – often resulting in growing nationalism and North-South binaries. As such, the European model has been seriously questioned (Holden 2012; Vollaard 2014). In the context of the simultaneity of these evolutions – of establishing an institutional basis for global EU actorness on the one hand and of the financial crisis undermining the European model on the other – this book studies the emergence of a collective European diplomacy. It covers the first four years of the EEAS' existence that Ashton refers to in the quote above, and engages with both 'internal and external factors' as well as with perceptions of 'what is being said about it' by different 'audiences'. With the last fieldwork phases concluded in March 2014, the book is limited to these four years and does not address the further evolution of the EEAS under Ashton's successor Federica Mogherini, whose tenure as High Representative/ Vice President commenced in November 2014.

Since the formal launch of the EEAS in December 2010, there has been a broad range of academic engagement with this new key institution of EU external relations (Carta 2012; Bátora 2013; Vanhoonacker and Pomorska 2013; Kuus

2014). Yet most of this research has been inward-looking, focusing on the EEAS' formation in Brussels and against the EU's established bodies of external relations. There is only limited engagement with the external dimension of building the new service, i.e. how collective diplomacy is enacted *abroad* and how it is perceived by the EU's external cooperation partners. This book addresses those two questions by focusing on the Kenyan capital of Nairobi as a key place where the EU's global actorness is enacted, in particular as regards its relations with Africa and role as a development actor.

The EU frequently claims a leading role in international development and as a key partner to developing countries (EU 2013). 'Development is at the heart of the EU's external action', the Commission postulates (EC 2014a) with a focus on sub-Saharan Africa. This focus on 'Development' applies to academic engagement with African-European relations, as research tends to concentrate on specific EU development policies and their macroeconomic effects, often neglecting other aspects beyond 'Development', such as diplomatic conduct, as well as the individuals and sites that enact these relations on both sides through daily interactions. I regard this gap as problematic because such research often lacks attention to detail and to social relations in particular. It also narrows down African-European relations to one policy field; a policy field often considered as reproducing hierarchical donor-recipient relationships (Escobar 1992; Chang 2004; Jackson 2005).

My approach seeks to go beyond this fixation with both Brussels and 'Development' by instead emphasising foreign policy and diplomacy in African-European relations and drawing attention to mundane practices and processes of how the EEAS is established *abroad* and how the EU's external cooperation partners perceive its *political* role. Figure I.1 (below) shows a map of EU Delegations in sub-Saharan Africa.[1] In total, the EU's diplomatic network comprises 139 Delegations, employing 5486 staff (1952 EEAS and 3534 Commission) around the world (EEAS 2015b). With 46 of those delegations in sub-Saharan Africa, employing a total of 1872 staff (610 EEAS and 1262 Commission) (ibid.), the distribution of the EU's diplomatic network there is higher than in other parts of the world, which underlines the importance of this macro-region to the *making* of EU diplomacy.

The Kenyan capital of Nairobi has been chosen for this study as one of Africa's political and economic centres and because of its pivotal role as a place of international development, politics and the enactment of African-European relations. Nairobi hosts numerous UN and development agencies. This omnipresent shadow of the 'development industry' (Hancock 1989) positions the city as an

1 The focus here is on sub-Saharan Africa. There are also EU Delegations in Morocco, Algeria, Tunisia, Libya and Egypt, however, in EU external relations, those countries are not included in the EU policy frameworks towards Africa, but towards the 'Mediterranean, Middle-East & the Gulf' (EEAS 2015a). Relatedly, also the term 'African-European relations' is used mostly here for referring to relations with sub-Saharan Africa.

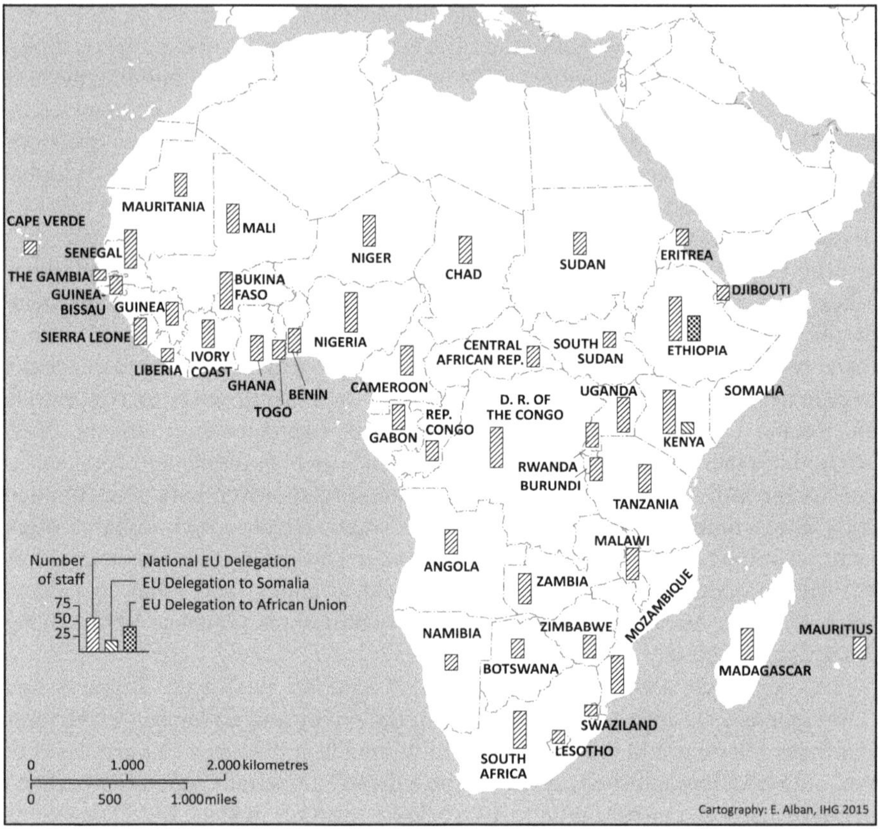

Figure I.1 Map of EU Delegations in sub-Saharan Africa (cartography by Elke Alban)

ideal case for studies that go *beyond* 'Development'. The processes and practices of forging a more active political role for the EU are particularly interesting as regards the question of how the EEAS succeeds (or not) in transforming the EU's role from that of a development actor into one of a diplomatic actor. Yet also more generally, Nairobi's location plays a central role for EU-African relations. Its international connectivity is widely documented in various areas, such as networks of NGOs (Taylor 2004) and international organisations (Taylor 2009), in terms of transport and mobility (Otiso et al. 2011; Pirie 2014), as an economic hub and potential world city (Scholvin and Draper 2012; Myers 2015) and as the general node of the EU's relations with East Africa, the Great Lakes Region and the Horn of Africa (Briet 2015).

Also on a continental scale, Nairobi is one of the key urban centres and Kenya is the political and economic power house in East Africa – not least through the importance of the port of Mombasa, but also as regards its political role in the region, such as in the Sudan conflicts and in Somalia (ONYANGO-OBBO 2012). The choice of Nairobi in this study is thus more than an engagement with one city or one country; it was deliberately chosen as a key site of African-European interaction that makes it possible to focus on the EU's *political* role in a 'Development' environment. Moreover, the EU has had a long and extensive engagement with Kenya. The Nairobi Delegation opened in 1976 and is now amongst the largest in the world, accredited to Kenya as well as to the United Nations Environment Programme (UNEP) and the UN Centre for Human Settlement (UN-Habitat), both of which are headquartered in Nairobi. The physical premises of the Nairobi Delegation also contain the EU Delegation to Somalia. The site of this fieldwork is therefore a key outpost of EU diplomacy of relevance much beyond the EU's relations to Kenya. It was also one of the external missions where the changes of the Lisbon Treaty had been implemented in a first trial phase before the official launch of the EEAS in December 2010 (Author interview, March 2012). This offered the opportunity to study practices and processes of establishing the new service earlier than in other locations. It simultaneously underlines the distinct role of Nairobi as a pivotal place of African-European interaction and of the Nairobi Delegation as a key factor in the making of EU diplomacy *abroad*.

As such this study is situated at the interface of European foreign and development policy. It is sensitive to wider historical and international contexts – such as colonialism, European civilian and normative power debates, the emergence of the Euro crisis, etc. – as well as to the local context of the specific situation in Kenya during the time of the research. The focus is on two closely-related aspects. First, this study develops an understanding of how political and diplomatic relations are coordinated and how 'Europe' positions itself collectively. It does so by examining the specific coordination processes between the EU Delegation and the member states' embassies in Nairobi. Second, it addresses the interaction with and the perceptions of the EU's external cooperation partners, i.e. the Kenyan government and other national and international organisations in Nairobi. It thus traces internal processes and practices of positioning the EU as *one* collective actor, on the one hand, and analyses perceptions and expectations of its external cooperation partners, on the other. Both aspects are approached in light of the simultaneous implementation of the Lisbon Treaty (and the EEAS) and the emergence of the Euro crisis – with the former resulting from a long process of designing an institutional basis for greater global influence of the EU and the latter severely impairing the EU's geopolitical appeal as a model for effective regional integration. In the above-mentioned speech, Ashton also displays a keen awareness of these evolutions when she says that the 'Lisbon arrangements for external policy […] was [*sic*] conceived in better economic times: so my job was to turn those aspirations into reality in the midst of the worst economic crisis the EU had ever faced' (EEAS 2014d). In light of these transformational processes, the twofold objective of this study is to understand *how* the EU is positioned and perceived as a geopolitical actor through the EEAS, a perspective that I attempt to achieve through a critical geopolitical lens.

A Critical Geopolitical Approach to EU Diplomacy

Since the late 1980s, the school of thought of critical geopolitics has been concerned with exploring how dominant discourses shape geopolitical agency and processes. Critical geopolitics questioned hitherto unchallenged assumptions about the constitution and workings of 'geopolitics', in particular the assumption that geopolitical organisation is determined exclusively in terms of territorial nation-states. As such it focused on discourse-centred deconstruction and analyses of geopolitical conditions and meaning. Geopolitics was thus to be 'critically re-conceptualized as a discursive practice' (Ó Tuathail and Agnew 1992, 192) and respective analyses be concerned with 'the politics of the production of global political space by dominant intellectuals, institutions, and practitioners of statecraft in practices that constitute "global politics"' (Ó Tuathail 1996, 185). Since the 1990s, critical geopolitics has evolved and broadened as both a methodological and conceptual lens for geopolitical inquiry which encompasses 'various ways of unpacking the tropes and epistemologies of dominant geographs and scriptings of political space' (Power and Campbell 2010, 244). A central element of critical geopolitical research has therefore been the analysis of dominant discourses (Müller 2010). After an initial focus on text, more recent critical geopolitical scholarship has moved to include also social practice into discourse analysis by using ethnographic and other research methods (Megoran 2006; Müller 2008; Kuus 2012). Employing such an extended understanding of discourse, Müller (2010, 13–14) posits that 'critical geopolitics does not *work* with discourse analysis as an instrument but it rather *is* discourse analysis'.

'Discourse' is thus considered holistically, taking into account that geopolitics is dependent not just on governmental texts and speeches but also on an expansive sphere of cultural texts, gestures, and practices, that can be loosely categorised in three distinct albeit interactive spheres: formal, practical and popular geopolitics (Ó Tuathail 1999; O'Loughlin, Ó Tuathail, and Kolossov 2005). *Formal* geopolitics refers to 'formalized theories and grand strategic visions of geopolitical intellectuals' (Ó Tuathail 1999, 113), developed, articulated and analysed predominantly by professional (geopolitical) thinkers in academia, think tanks and other research institutions. *Practical* geopolitics describes the 'everyday practice of statecraft' (ibid., 111) – not only in the form of the implementation of geopolitical action through politicians, civil servants and other public office holders, but also in the form of official geopolitical reasoning through public institutions and channels. *Popular* geopolitics refers to popular culture and its role in the production of geopolitical 'commonsensical' knowledge, including literature, the visual arts, movies, television, newspapers and, of course, online media sources. This sphere forms an important aspect of critical geopolitical analysis because in a democratic polity foreign policy choices require, to some degree at least, public legitimation. Nonetheless it would be a mistake to reduce this to a unidirectional relationship, as formal and practical geopolitical agents are themselves shaped through popular cultures (Dittmer and Dodds 2008).

In this combination of formal, practical and popular spheres of geopolitical analysis, critical geopolitics opened the door for a systematic, conceptual engagement with social and discursive dimensions of 'making' geopolitics, including stories, emotions, perceptions, desires, etc. Dominique Moïsi's (2009) insightful account of the 'geopolitics of emotion' lays out how a particular culture's geopolitical identity and role depends on and is fundamentally shaped by (one's own and external) perceptions and emotions about a culture/polity and how, consequently, these emotions influence geopolitical constellations and interactions with others (see also the classic works by Fanon 1961; Said 1978). The geopolitical narrative told about a culture or a polity – both by itself and by others – is therefore a crucial component for the internal functioning of this culture and its geopolitical positioning. Yet, only relatively recently has critical geopolitics been applied to the study of the EU (Sidaway 2006; Kuus 2007; Bachmann and Sidaway 2009b; Bialasiewicz et al. 2009; Bialasiewicz 2011; Kuus 2011b, 2011c; Scott 2011; Rovnyi and Bachmann 2012; Moisio et al. 2013). This is surprising given the permeability of ideas and personnel between different formal, practical and popular agents in both the actual processes of European integration as well as the accompanying discourses.

Whilst the three geopolitical forms (formal, practical and popular) are always interlinked and difficult to regard isolated from each other (Ciuta and Klinke 2010), the empirical focus of this book is on *practical* agents, key African and European diplomats and officials enacting African-European relations. Following Toal's call for a grounded critical geopolitics (Dahlman and Ó Tuathail 2005; Ó Tuathail 2010), my analysis traces these agents and seeks to understand their interaction. To achieve this, it is based on extensive in-depth fieldwork in the Kenyan capital of Nairobi. A critical geopolitical approach that I found useful is the one laid out by John Agnew and Stuart Corbridge in their classic *Mastering Space*. By geopolitics, they understand

> the division of global space by institutions (states, firms, social movements, international organizations, armed forces, terrorist groups, etc.) into discrete territories and spheres of political-economic influence through which the international political economy is regulated materially and represented intellectually as a natural order of 'developed' and 'underdeveloped', 'friendly' and 'threatening' areas. It is that set of socially constructed, rather than naturally given, practices and ideas through which the international political economy is realized geographically. [...] We share with an emerging school of 'critical geopolitics' the view that geopolitics is implicit in both the practice of and writing about all types of international relations (Agnew and Corbridge 1995, 4–5)

Although a plethora of critical geopolitical writings have emerged since the late 1980s (Jones and Sage 2010; Kuus 2010a; Dodds, Kuus, and Sharp 2013; Moisio 2015), *Mastering Space* still functions as a useful point of departure for critical geopolitical analysis. Also for this book, I draw on Agnew and Corbridge's approach. Yet I contest their use of a terminology of 'divisions of global space' cited at the beginning of this quote. Instead, I prefer to think in terms of an ordering, or the organisation or regulation, of global space. Whilst I do not intend to deny the existence of ongoing, often hierarchical, divisions between Africa and Europe, my focus here is not on divisions but on interactions. It is on questions of how the EU positions itself as a diplomatic actor in Kenya, how different European agents interact with each other and with Kenyan and international partners, and how the EU and the interactions with the EU are seen by its cooperation partners. Divisions, rooted in Europe's imperial past (and present) as well as in contemporary and unequal power relations, shape these interactions and perceptions. As such divisions (and perceptions thereof) influence my work, yet the focus is on how 'global space' is organised and regulated through practices and processes of *interaction* between different agents. It is on how the divisions are negotiated and dealt with, and not on the divisions *per se*.

In the same light, I also contest Agnew and Corbridge's use of terminology in which the divisions of global space are seen in terms of 'discrete territories and spheres of political-economic influence'. My interest is not in territories. 'Spheres of influence', on the other hand, play a role in my work – in the sense of a collective European influence in Africa. Yet this is also not at the heart of my study. I regard it as problematic to pitch such spheres of influence since they inevitably raise questions of geopolitical rivalry. While I touch on the question of how the growing Chinese role in Africa influences African-European relations, I am not interested in the emergence of a possible Chinese-European battle for a sphere of influence in Africa. Instead I try to understand and demonstrate how a collective European diplomacy emerges in one of the key sites of global North-South relations. This is thus not about geopolitical rivalry or spheres of influence: it is about the transformation of EU geopolitical agency.

Beyond these two aspects, Agnew and Corbridge's understanding of geopolitics provides an important grounding for the approach employed here. Their focus beyond the nation state and on 'institutions (states, firms, social movements, international organizations […])' is important for my interest in studying the geopolitical agency of the European Union. Equally important is their concern with 'political-economic influence' and the regulation and 'intellectual representation' of the 'international political economy'. My focus is more political than economic, yet the economic component is essential for understanding external perceptions of the EU in Africa. Perceptions are differentiated along a binary of the appeal of the political model of regional integration and civilian power on the one hand and resentment against the EU's economic power and (neo-)imperial practices on the other (Bachmann 2013a). These are two sides of the same coin that together shape external perceptions of the EU and the positioning of the EU as a collective actor. It is not only political decisions and factors – such as the Lisbon Treaty or the establishment of the EEAS – or geopolitical evolutions – such as an increasing Chinese engagement in Africa – that determine the EU's external agency. Its international role is tied into wider economic contexts, such as trade policy, economic relations and the emergence of the financial crisis in 2008. Agnew and Corbridge's understanding of geopolitics allows for capturing both political and economic factors and constellations of the organisation of global space and international politics.

Most important in their definition, however, is the emphasis on '*socially constructed*, rather than naturally given, practices and ideas through which the international political economy is realized geographically'. Geopolitics in this sense is thus not static in time or in space but 'bears witness to the irredeemable plurality of space and the multiplicity of possible political constructions of space' (Ó Tuathail and Dalby 1998, 2–3). The focus on social constructions and the malleability of 'geopolitics' is, of course, a central tenet of critical geopolitics more generally. As such it is also central to the research approach employed here. My focus is on the *dynamic* social construction of EU diplomacy, not on static legal documents (such as the Lisbon Treaty) or structures that only provide the framework for agency. Yet the understanding of geopolitics employed here is broader. The prefix 'geo' has multiple meanings. On the one hand, I use it in an often criticised[2] but nevertheless common meaning of the prefix in the sense of 'global' or, rather, as introducing a wider international perspective on political and economic processes – which Agnew and Corbridge refer to as 'all types of international relations'. On the other hand, it refers to the '*geographic* realization of the international political economy'. In contrast to global politics, the understanding of geopolitics employed here accounts for a variety of forms of interplay.

Figure I.2 (below) provides a sketched abstraction of these interplays and of how I make use of critical geopolitical reasoning in this book. The proposed approach seeks to capture how social construction is historically and spatially conditioned, how it is politically and economically framed, how it is locally and internationally embedded and how it is both materially and representationally constituted. It views geopolitics in terms of 'space-time' (Massey 1994) and as processes conditioned through historical development and the socio-political nature of space. It also takes account of the interdependence of political and economic framings of geopolitics in terms of interests, capabilities and resulting actorness. Simultaneously, geopolitics is approached here as 'localised' (Dahlman and Ó Tuathail 2005), as embedded in both the local context of the specific research site as well as in wider international (power) structures and networks. And finally, I emphasise the coexistence of representational and material factors of geopolitics, i.e. geopolitics as constituted both by imaginations, perceptions, experiences, expectations and narratives as well as by materiality of 'power', capital, physical infrastructure, etc. While I emphasise the importance of all of these factors and their interplay in critical geopolitical analysis, the focus of this

2 Elden argues that this is an unnecessary reduction of the meaning of the term geopolitics and instead suggests a wider meaning of 'geo' in terms of 'land', 'earth', 'world' (Elden 2013, 49).

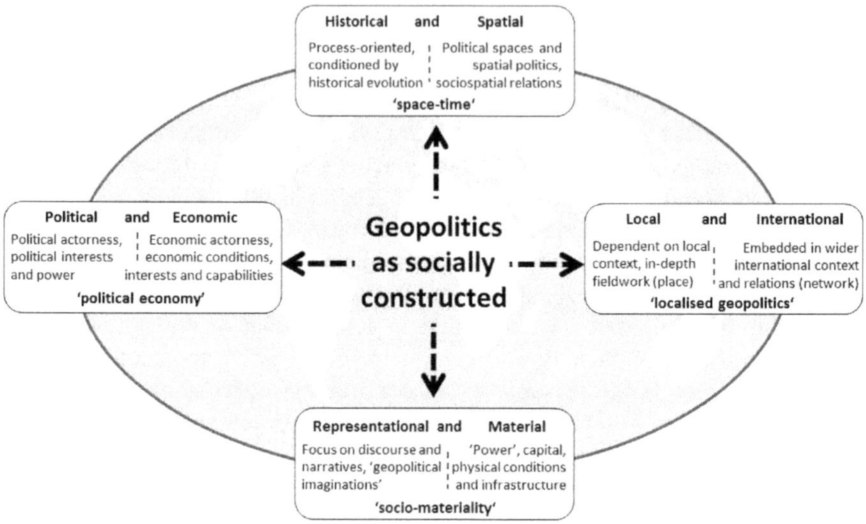

Figure I.2 Abstraction of the critical geopolitical approach of this book

book remains on social construction. It is also more concerned with political framings than with economic ones, more interested in local than international embedment and focuses more on representations than on materiality. It is in such a context that this book provides a grounded geopolitical analysis of EU agency in Kenya, focusing on perceptions and interactions of key individuals enacting African-European relations.

Through such a critical geopolitical approach, I seek to contribute to and advance a range of different debates in human geography, European studies, and international relations. This book primarily speaks to wider debates on Africa-EU relations and the EU's role as a global actor (Bretherton and Vogler 2006; Manners 2010; Bialasiewicz 2011; Adebajo and Whiteman 2012; Telò and Ponjaert 2013). Yet it has the ambition to accentuate the importance of a detailed grounded geopolitical analysis of how the EU's external agency is enacted in a variety of different empirical settings. The key area of interest is not development policy, but foreign policy and diplomacy. As part of wider institutional evolutions and increasing geopolitical ambitions during the 2000s, these policy fields became subject to EU-level politics (Duke 2009; Benson-Rea and Shore 2012; Carta 2012; Kuus 2014; Smith, Keukeleire, and Vanhoonacker 2014). Through the Lisbon Treaty, intergovernmental coordination of foreign policy and diplomacy became physically institutionalised in the form of the EEAS. Despite not being supranational, the new service invokes supranationality. Nevertheless, collective diplomacy remains a policy field 'in which authority is dispersed, interests are diffuse and often conflicting, and international political identity is blurred' (Carta

2012, 85). In addition, as part of these wider debates on the EU's 'international political identity', this book focuses on external perceptions and expectations of the EU in order to develop a more reflexive understanding of its global actorness. This is particularly important as regards Europe's relations with regions where relations have long been hierarchical (for instance through colonialism or donor-recipient aid). As such it speaks to a set of prior literature that has studied how the EU is seen from the outside (Lucarelli and Fioramonti 2010; Bachmann 2013a; Beauguitte and Didelon-Loiseau 2013; Chaban et al. 2013; Holland and Chaban 2014; Larsen 2014).

To both debates – the EU as a global actor and perceptions of the EU from the outside – this research adds a critical geopolitical take, emphasising the need to study how EU diplomacy and perceptions thereof emerge in processes of social construction, how they are historically and spatially conditioned, politically and economically framed and, of course, how they are locally and internationally embedded. The study of diplomacy through such a critical geopolitical lens zooms in on 'dominant intellectuals, institutions, and practitioners of statecraft' and the practices through which those constitute global politics (Ó Tuathail 1996, 185). It conceives of diplomats as 'central agents of geopolitics' (Kuus 2014, 34) and of diplomacy as 'an integrating mechanism for what is now emerging as a global polity; [a mechanism that is] as significant as trade and war' (Neumann 2012, 1–2). In so doing, the approach employed here essentially builds on two key trends in critical geopolitical research on diplomacy. First, it builds on the 'gradual weakening of state-centrism in the social science' (Kuus 2014, 34) and, by extension, attention to the geopolitical agency of non-state actors in the study of diplomacy (McConnell, Moreau, and Dittmer 2012; Bachmann 2013c). My work contributes to this trend by focusing on the emergence of a collective European diplomacy and by studying the practices of establishing its key institution, the EEAS, *abroad*. Second, it builds on a trend that lays a 'more sustained focus on geopolitical practices rather than scripts' (Kuus 2014, 36) and the everyday practices of enacting diplomacy and geopolitics (Mathiason 2007; Müller 2008, 2012; Neumann 2012). My work also adds to this trend by studying the mundane practices and processes of enacting the EU's *political* (not development) role in Nairobi through in-depth fieldwork, institutional ethnography and active cooperation with African and European diplomats.

Methodological Framing

The methodological approach employed for this book centres around interviews and participant observation. Since 2006 I have conducted seven fieldtrips to the East African region with a combined duration of 15 months of fieldwork, mostly in Nairobi. An earlier project studied collective European development policy towards the African Union (AU), the East African Community (EAC) and the Republic of Kenya between 2006 and 2009 (including 55 elite interviews and six months of participant observation with a European development organisation). Some of the empirical data of this earlier project informs the analysis presented here, yet this book is primarily based on research conducted in Nairobi and Brussels as part of the *EuroGaps* project between 2010 and 2014.[3] The methodological centrepiece of this research were 58 semi-structured interviews conducted with 56 key individuals involved in the establishment of the EEAS as a collective diplomatic service and in Kenyan-European interaction on the *political* level. Informants included senior officials (Ambassadors, Deputy Ambassadors, Directors, etc.) from the EEAS headquarters in Brussels as well as from the EU Delegation, EU member states, the Kenyan government and the NGO sector in Nairobi. For reasons of confidentiality, the names and exact functions of the informants are kept anonymous throughout the text. However, I do indicate their affiliation, for instance, 'member state diplomat', 'EU diplomat', 'Kenyan official', etc.

In addition to conducting interviews, I held an affiliation with the Delegation of the European Union to the Republic of Kenya in Nairobi. To conduct this institutional ethnography, I was based in the Delegation as research associate for a period of three months. By the time of my first fieldtrip to Nairobi in the context of this study, the EEAS had not yet been formally launched as an institution. Access to people knowledgeable about the new service, in particular in a sensitive field such as diplomacy, was therefore difficult. The affiliation with the EU Delegation in Nairobi was thus an invaluable asset to my research as it gave me access to the interior of the key institution charged with building the EEAS in Nairobi. Furthermore, it allowed me to study the sites of diplomatic conduct between the different European partners and so develop an understanding of the social dynamics within and between them in one of the largest EU Delegations in the world. I also held affiliations with the Kenyan Foreign Ministry's *Foreign Service Institute* and the University of Nairobi's *Institute of Diplomacy and International Studies* (IDIS) – the two key training facilities for diplomats and staff of international organisations in the East African region (the former a governmental one, the latter an academic one). Through these affiliations I was exposed to a wide range of high-level informants in Kenyan and European institutions in Nairobi with whom I interacted on a frequent basis both formally (in the form of interviews or through my affiliations) and informally (at receptions, dinners or through plain conversation).

In employing multiple methods, I seek to '*complement* discursive and representational studies' (Megoran 2006, 627, emphasis added) on the EU's geopolitical identity, such as my own prior work on EU geopolitical discourse (Bachmann and Sidaway 2009b). Instead of merely deconstructing such discourses per se, I develop new insights by studying how collective diplomacy is made – both in relation to such discourses and beyond. In line with the critical

3 Entitled *EuroGaps: External Relations and External Perceptions of the EU in sub-Saharan Africa and the Black Sea Region*, and funded by the German Ministry for Education and Research (BMBF) between 2010 and 2014 (see www.eurogaps.eu).

geopolitical focus on social construction outlined above, my research thus centres on the practices and processes of how key individuals interact with each other and perceive the EU – and thereby construct or ascribe actorness to it.

Chapter Outline

This book is comprised of two parts containing four chapters each, plus an introduction and a conclusion. The first part lays out the wider background as well as the conceptual and methodological framework of the study. The second part is largely empirical, presenting how the EEAS is positioned and perceived in Kenya as well as discussing wider expectations and perceptions of the EU's geopolitical role.

Chapter 1 provides an overview of the processes of European integration and the formal political frameworks through which the EU emerged as a key geopolitical actor. It traces the integration steps in a historical perspective, focusing on the integration of collective external agency and in particular on the EU's relations with developing countries. Chapter 1 also lays out the formal processes of establishing the European External Action Service and the role of the *High Representative of the Union for Foreign Affairs and Security Policy/Vice President of the European Commission* (HR/VP). It outlines the difficulties in defining the HR/VP's (and by extension the EEAS's) precise function and role in and against the EU's other external relations instruments. Drawing on interviews with EU diplomats, Chapter 1 also discusses some of the problems that surrounded the early phase of establishing the EEAS in practice. It introduces the term 'two-step diplomacy' to describe the peculiarity of diplomacy of a heterogeneous and composite actor like the EU, i.e. the process of having to initially negotiate a collective position amongst EU partners before this position can then enter 'traditional' diplomatic negotiations with external partners.

Chapter 2 traces the evolution of positioning the EU as a global and development actor in terms of geopolitical discourse. It briefly outlines how European integration has been accompanied and influenced by civilian and normative power debates whereby the evolving European polity has often been idealised as a geopolitical model for regional integration to be emulated elsewhere. In addition, Chapter 2 illustrates how the EU's cooperation partners in East Africa indeed perceive(d) the EU as such a model and expect(ed) the EU to assist African regional organisations in their integration efforts. It argues that a key factor in legitimising a global role for the EU has been the appeal of its model of regional integration. The Lisbon Treaty, as a key moment in the EU's institutional evolution, was a crucial step in consolidating this 'model' path of European integration. With it, the EEAS was an equally crucial step in adapting the EU's external role to its internal level of integration and positioning it globally as a more cohesive and influential (geo) political entity. However, the coming-into-effect of both the Lisbon Treaty and the EEAS coincided with the eruption of the crisis in the Eurozone. I argue that, because of the crisis, the functioning of Europe's model has been seriously called into question.

Chapter 3 introduces a spatial approach to the dominant geopolitical discourses presented in the previous chapter. It draws on key geographical texts to help emphasise the spatiality of political agency and the political nature of space. I introduce the term 'spaces of interaction' to describe a broad framework that governs the interaction of various political, economic and social actors. I argue that the spaces of interaction that have emerged through the European integration process are mostly regulated and institutionalised. These spaces of interaction – guided by civilian and normative power ideas – are also being projected in and through the EU's external relations. Moreover, this chapter provides a brief geographical interpretation of the literature on Europeanisation. Whilst Europeanisation is a well-established field of research on EU-internal processes, its external dimension is underdeveloped. This applies to processes of Europeanisation in interactions of Delegations and member states embassies abroad. However, it applies even more to the question of how processes of Europeanisation affect interaction with external cooperation partners, such as the Kenyan government, local NGOs or third countries. Through a critical geopolitical lens, I differentiate between *Europeanisation abroad* and *external Europeanisation*.

Chapter 4 reflects on the research process of studying diplomacy as a social process and an institution in the making. I predominantly employ qualitative research methods such as participant observation, interviews and focus group discussions. In-depth and long-term fieldwork was essential to the research process, which comprised affiliations with key institutions involved in the study, such as the EU Delegation in Nairobi, the Kenyan Foreign Ministry and the University of Nairobi. My research included both more formal methods and the day-to-day pragmatism of immersing oneself as a researcher in the environment and the situation to be studied. In this chapter, I emphasise the importance of studying the sites of professional and social interaction as well as the social dynamics of these places to understand how decisions are made and how networks, through which EU external relations operate, are being forged and evolve. Such in-depth fieldwork allows for the development of an understanding not only of the formal relations, but also of the informal and social dynamics that enact formal relations.

Chapter 5 lays out the processes of establishing the EEAS between the different European partners, i.e. the Delegation and the member states, in Nairobi. It describes how political and diplomatic relations are coordinated and how 'Europe' positions itself collectively. It adds a more practical view to the formal integration steps described in Chapter 1 and discusses how collective diplomacy becomes operationalised. In terms of the EU's two-step diplomacy (see Chapter 1), it thus focuses on the first step, i.e. the coordination mechanism and negotiations between the different European partners, and describes processes of *Europeanisation abroad* (see Chapter 3). This includes different EU institutions, such as the Commission and the EEAS, but also relations between the member states and EU institutions. Moreover, this chapter also points to the difficult transition of EU institutions in transforming the Commission's traditional role as a technical implementer of development policies into a more active diplomatic actor of the European Union. It thereby argues that the EU's preference for regulation and institutionalisation of spaces of interaction (see Chapter 3) is ill-equipped to deal with the challenges of diplomacy which is more dependent on (political) sensitivities, personal relations and trust.

Chapter 6 addresses the interaction with and the perceptions of the EU's external cooperation partners, such as the Kenyan government and other organisations in Nairobi. It discusses how the EEAS is achieving recognition as the key European foreign policy institution amongst its cooperation partners in Kenya. It focuses on the second step of diplomacy (see Chapter 1) and the related processes of *external Europeanisation*. Chapter 6 thus adds perceptions of the EU's external cooperation partners as regards the EU's transition from a development to a political actor (see Chapter 5) and with respect to idealised policy fields of multilateral or bilateral interaction. It also discusses the specific political situation in Kenya around the 2013 elections and the cases of President Kenyatta and Vice-President Ruto at the International Criminal Court (ICC) in The Hague following their alleged involvement in the post-election violence of 2008. In this context, this chapter illustrates how this specific situation created an environment that contributed to shifting collective European agency away from institutionalised and regulated policy conduct and towards personalised interaction – a process through which the EU Head of Delegation in Nairobi emerged as the key figure of European diplomacy.

Chapter 7 focuses on *expectations* of the EU's role as a global and development actor. It teases out different expectations depending on the respective background and institutional affiliation of the informant or source of information. As such, Chapter 7 is a presentation of the very subjective opinions of the informants about what role Europe *should* be playing. However, through this subjectivity it illustrates how expectations vary depending on affiliation, background and viewpoint. It addresses three types of expectations: a) informants' expectations of the EEAS as the new institution of the EU's external agency; b) expectations of the EU's role in Kenya and the wider East Africa region; and c) expectations of the EU's global role. The results (based on research conducted from 2010 to 2014) are set in context with the results of an earlier research project (drawing on research data from 2006–2009), showing both changes and persistence with regard to expectations about the EU. The chapter concludes by highlighting how African cooperation partners expect the EU to play the role of a partner without paternalisation.

Chapter 8 addresses informants' general perceptions of the EU's wider geopolitical role as a global and development actor. It starts by reviewing how informants still consider regional integration as the inherently 'European thing' and discusses how changing geopolitical circumstances have influenced how the EU is seen in a global context. The chapter then contrasts perceptions of the EU as a (geoeconomic) power and as a (geopolitical) model, before laying out how the EU is perceived as an actor in international development. Moreover, it also illustrates areas and policy fields in which the EU is seen in a particularly positive or in a particularly negative context. As such this chapter is much more general than the previous ones. Instead of analysing how the EEAS is built amongst European partners (Chapter 5), interacts with and is perceived by its Kenyan partners (Chapter 6), or what is expected from the EU and its new service in Kenya (Chapter 7), this chapter is concerned with general perceptions and imaginings of the EU's identity and role in the world. It concludes that the regional integration aspect still features prominently in perceptions of the EU but also that, in light of the European debt crisis and wider geopolitical transformations, this does not directly translate any more into expectations for promoting the European model abroad.

The Conclusion reviews some of the key points made in this book and includes my personal, normative outlook on the EU's future role. I critically discuss the backgrounds to the difficulties of transforming the EU from a technical development implementer to a political and diplomatic agent, arguing that the personalised conduct of diplomacy stands in contrast to the EU's established regulated and standardised *modi operandi*. In Kenya, it was through the special situation of the ICC cases that collective diplomacy emerged. In this context, I also emphasise the value of the critical geopolitical approach employed here to the study of diplomacy and adapt the schematic abstraction of the approach offered in the introduction to the specific case studied here. Finally, I discuss the continued relevance of the regional integration narrative to perceptions of the EU and highlight the importance of taking external perceptions of the EU seriously. In particular as regards African-European relations, perceptions of past and present imperialism persist widely. Rather than evaluate such perceptions, this study emphasises their existence and impact on these relations. In so doing, I argue that positioning the EU as a more reticent geopolitical actor is not necessarily to its disadvantage and should not be regarded as a failure of the European project.

PART I
External Relations at a Critical Moment

Chapter 1
Instruments of Conduct for European External Relations

Over several decades of integration, the EU has taken multiple steps to increase its global political role. This book provides a detailed snapshot of the most recent phase of such attempts to grow, i.e. the making of diplomatic agency through the EEAS. Yet before this phase, the European integration process was accompanied by a long institutional evolution that eventually produced the EEAS. In this chapter, I trace this evolution. I review the positioning of the EU as a global and development actor in terms of policy frameworks and external relations instruments, such as the European Political Cooperation (EPC) in the early 1970s, the Common Foreign and Security Policy (CFSP) as the second pillar of the European Union between 1993 and 2009 and, eventually, the European External Action Service (EEAS). Unlike those previous measures to harmonise European foreign policy, the EEAS is a legal, separate and independent instrument with a joint mandate from the European Commission and the Member States (through the European Council) to represent the Union on all external relations. It thereby adds an institutional component to existing forms of cooperation of the member states in the fields of diplomacy and foreign policy.

An Emerging Collective Diplomacy through Development Policy

Institutional arrangements for the collective conduct of external relations as part of the European integration process date back to the 1950s. Yet the early steps of integration in particular were predominantly concerned with internal issues. The contract for the European Coal and Steel Community (ECSC), signed on 18 April 1951, regulated collective control over the French and German coal and steel industries under a collective High Authority and created a common market for these two key products of Western European reconstruction. It is commonly considered as the first legally binding contract of supranational integration in the history of international relations. In addition to France and the Federal Republic of Germany[1] (FRG), Italy and the Benelux-countries signed the contract; the UK did not, based on a decision not to accept the principle of supranationality. While the improving relations between France and Germany led to a range of bilateral

1 In the following referred to as Germany; in contrast the German Democratic Republic (GDR) will be referred to by its full name.

agreements in economic and cultural areas the other partners of the ECSC, Italy and the Benelux countries, were wary about this growing French-German bilateralism and pushed for a more inclusive multilateral approach – leading to the Treaty of Rome, signed on 25 March 1957 (Thiemeyer 2010, 50–52). The treaty established the European Economic Community (EEC) and transformed the ECSC's High Authority to the European Commission. The agreements signed in Rome were the first to include sections on external relations, focusing in particular on the then colonies of the signatories. Mainly due to French and Belgium pressure, the treaty included clauses demanding other EEC members to make financial provisions for the (former) colonies (Dialer 2007, 45). Holden (2009, 126) points out how

> as the French Union faded France promoted a structural relationship between its ex-colonies and the EU as a part (and only one part) of its continued influence. The overlap between the origins of EU development policy and European colonial projects gives ready ammunition to critics of the EU's role in Africa, but this historical conjuncture is not in itself proof of neo-imperialism.

The collective relations with Africa took priority in collectivising the EEC's external relations. The early agreements in the Treaty of Rome led to the establishment of the European Development Fund (EDF) in 1958 and the Yaoundé Convention between the six countries of the EEC and the 18 countries of the group of Associated African States and Madagascar (AASM)[2] in 1963. During the same time, and through its collective approach to development policy, the European Commission also took initial steps towards developing a diplomatic network by establishing first delegations. The Commission's Directorate General for Development (DG VIII) had been implementing collective aid programmes in overseas territories from the early 1960s. As such its 'outpost model [...] became a template for subsequent Commission efforts to be present in third countries' (Carta 2012, 55). These delegations pursued less of a political-diplomatic mission, for, known as *contrôleurs techniques du Fond Européen de Développement*, they functioned more as implementing agencies for development policy. Simultaneously, and in addition to DG VIII, the Commission's Directorate General for External Relations (DG I) also opened delegations, mainly with the key commercial partners. The first of these delegations to enjoy full diplomatic status opened in 1956 in London – still as a representative of the High Authority of the ECSC (Carta 2012, 56–7).

In the mid-1970s then, the Yaoundé Convention was replaced by the Lomé Convention between the EC and the 43 countries of the African, Caribbean and Pacific Group (ACP). It was praised as a turning point in global North-South relations in a sense that it moved EC-ACP relations away from dependent 'association agreements' to 'partnership' and 'cooperation' (Dialer 2007, 50–51).

---

2 Benin, Burkina Faso, Burundi, United Republic of Cameroon, Central African Republic, Chad, Congo-Brazzaville, Congo-Kinshasa, Côte d'Ivoire, Gabon, Madagascar, Mali, Mauritania, Niger, Rwanda, Senegal, Somalia, Togo.

Over the next 25 years, the Lomé Conventions were renegotiated and renewed every five years (Lomé I–IV); by the time it was replaced by the Cotonou Agreement in 2000, it had developed into a comprehensive and long-term interaction framework including the then 15 countries of the European Union and 71 ACP-states. The Lomé Conventions were characterised by aid and trade preferences for former colonies of European countries and, as such, were limited in both geographical and policy scope. When the foundation of the European Union in 1993 established the basis for a Common Foreign and Security Policy (CFSP) the EU 'wanted to evolve into a global actor and EC external assistance became an ideal tool. The EC become actively involved in all regions of the developing world, but this generated an overstretched and fragmented policy'. The Treaty of Maastricht thereby laid the 'foundations for a change of directions, establishing the principles of co-ordination and complementarity and granting the European Commission the role of promoter of aid co-ordination' (Carbone 2007, 58–9).

European Political Cooperation (EPC) and the Common Foreign and Security Policy (CFSP)

The first enlargement in 1973, however, not only brought about changes in the design of collective development policy but also a shift to a more collective foreign policy and diplomacy. Already a few years earlier, in 1970, the Community of the Six initiated an institutionalised attempt to harmonise foreign policy in the form of the European Political Cooperation (EPC). These attempts lead to what Wessels (1982, 4–6) described as a 'coordination reflex' that had been injected into the diplomatic behaviour of EC member states so that increasingly national policy positions were developed only after consultations with European partners. Such adaptations of national policy making to the processes of European integration are now widely elaborated in the literature on Europeanisation (see Chapter 3 and Graziano and Vink 2008; Jones and Clark 2010; Rovnyi and Bachmann 2012; Wong and Hill 2012; Moisio et al. 2013). With regard to external relations these processes can be described as

> a transformation in the way in which national foreign policies are constructed, in the ways in which professional roles are defined and pursued and in the consequent internalisation of norms and expectations arising from a complex system of collective European policy making. (Tonra 2000, 229)

In the context of such external dimensions of Europeanisation, Wong refers to the notion of 'coordination reflex' and points to 'a *réflexe communautaire* [that] becomes the norm rather than the exception' (Wong 2008, 323). The early steps for this reflex were laid in the 1970s with 'the foreign ministers work[ing] together 'intergovernmentally' on the basis of non-binding agreements that do not provide for formal or permanent institutions' but with the general aim of putting 'the Nine into a position whereby they can speak with one voice on international questions'. The EPC was thus a 'structure which favours and facilitates such approximation of points of view but does not necessarily produce them' (Wessels 1982, 1–3). In contrast to the measures undertaken in the 1960s and 1970s in the area of collective development policy, the EPC and the integration of foreign policy was strictly intergovernmental and 'excluded any contamination by European institutions' (Carta 2012, 62), such as the Commission or its delegations. In diplomacy, the EC, according to Allen and Smith (1990, 23), was 'far from being a unified actor. Rather it is a flexible and disaggregated series of patterns, arrangements and institutions which expresses a collective yet pluralistic identity, and of which others are increasingly aware'.

For Commission delegations, the entry of Spain and Portugal in 1986 brought about major changes as the Commission's delegation network was extended from its previous focus on Africa (or, more generally, former French and British colonies) to Central and South America and the Mediterranean. Moreover, the new head of the Directorate-General for External Relations, the Spaniard Ángel Viñas, not only pushed for a geographical extension of the network but also for an extension of delegations' competences beyond development to include political, commercial and diplomatic aspects. This expansion of geographical reach and competences through the Commission, and the homogenisation of its delegations as 'quasi-diplomatic missions' was met with limited enthusiasm by most of the member states (Carta 2012, 63). However, the Commission's supranational push into external relations beyond development certainly contributed to an increased concern with the external dimension in the intensifying integration debates of the late 1980s. The rivalry between Commission and member states in third countries might well have injected a dynamic to member states towards intergovernmental integration of foreign policy as part of the negotiations that eventually led to the Treaty of Maastricht and the creation of the second pillar of the EU, the Common Foreign and Security Policy (CFSP). A summarising overview of the various EU development and foreign policy frameworks discussed in this chapter is shown in Figure 1.1.

In the beginning, the institutional structure of operating the CFSP was similar to that of the EPC. Yet, negotiation and decision-making increasingly took place in Brussels – not necessarily supranationally, but by 'institutions and people based in Brussels rather than in national capitals' (Smith 2008, 38). The CFSP was more 'Brusselised' (ibid.) than the EPC. Also the Commission was more involved, not least because the CFSP budget was integrated and administered by the Commission. Nevertheless, decision-making remained firmly intergovernmental.

> The European Council set the broad guidelines for CFSP; the Council of Foreign Ministers implemented them, The Council's work with respect to CFSP items was prepared by the Political Committee (composed of the political affairs directors of the national foreign ministries). The Political Committee relied on the work of European Correspondents (foreign ministry officials) and working

Treaty	Development Framework (within the EDF)	Foreign Policy Framework
Rome (1957)	**Yaoundé (1963):** First association agreements with former (mostly French and Belgium) colonies under the EDF, hierarchical "aid"	**EPC (1970):** strictly intergovernmental attempt for exchange and coordination of foreign policy
Maastricht (1993)	**Lomé (1975):** Inclusion of former British colonies, rhetorical reorientation towards "partnership" and "cooperation"	**CFSP (1993):** intergovernmental, but more institutionalised and 'Brusselised', Commission participation
Amsterdam (1999)	**Cotonou (2000):** Comprehensive interaction framework between the Commission and ACP-countries, covering development, trade and other areas, rhetorical focus on "partnership"	High Representative of the CFSP
Lisbon (2009)		**EEAS (2010):** legal institution for the conduct of all EU external relations, mandated by Commission and Council, also political mandate

Figure 1.1 Overview of selected EU development and foreign policy frameworks

> groups. The European Commission was fully associated. The Parliament was informed of CFSP issues and could make recommendations, but its views did not have to be incorporated into CFSP decisions. The CFSP did not fall under ECJ jurisdiction. (Smith 2008, 38)

The Maastricht Treaty also allowed for Commission participation in CFSP meetings and closer cooperation with the respective national embassies, in particular that of the incumbent holding the rotating presidency which also represented the CFSP externally. This constituted a 'system of external relations, a collective enterprise which allows its members to pursue partly separate partly common international actions' (Hill and Wallace 1996, 5). In the one and a half decades following the Maastricht Treaty a number of changes have been put into place regarding the conduct of external relations. Most notable was the installation of a High Representative for the CFSP and a Policy Planning and Early Warning Unit in her/his support through the Amsterdam Treaty. The unit, albeit operating largely in the background, was an interesting experiment in attempting to bridge the different institutional camps that made up collective foreign policy. It was established within the Council Secretariat and staffed by Commission and national officials, but was dependent on the information shared by the member states (Smith 2008, 30–31). As such it functioned as a test run for the establishment of the European External Action Service (EEAS) a decade later. The contractual arrangements for the EEAS and its head, the High Representative of the Union for Foreign Affairs and Security Policy/Vice President of the European Commission (HR/VP), were first developed by the Convention on the Future of Europe. Headed by former French President Valéry Giscard d'Estaing, the convention drafted a constitutional treaty for the EU that served as the basis for the Lisbon Treaty, which eventually created the constitutional foundation for the EEAS and the HR/VP when it came into effect on 1 December 2009.

The EEAS and the HR/VP

European foreign policy before the Lisbon Treaty remained anchored in the principle of separation between the supranational external functions of the Commission and the intergovernmental ones of the member states. The former were predominantly restricted to areas of Commission competence, such as trade, fisheries, agriculture and development. Without a mandate for the Commission, collective foreign policy was coordinated, articulated and conducted by the country holding the rotating presidency. The Lisbon Treaty sought to improve harmonisation between all the Union's external actions – including, of course, those of the member states. Part of this ambition was to combine the former positions of External Relations Commissioner and High Representative of the CFSP in one post, that of the HR/VP. In general, the key motivation for creating this new post was to 'better ensure continuity, consistency and visibility for the EU in external affairs' (Cameron 2011, 239). Howorth identifies three main reasons for creating the 'double-hatted' post of the HR/VP.

> The first was political: the widely perceived desirability of ever greater coordination of the foreign and security policies of the EU's twenty-seven member states. The second was operational: the need for synergies, on the ground, between the main thrusts of CFSP and CSDP – trade, development aid, humanitarian assistance, and crisis management. The third was institutional: the growing recognition by member states that effective international action on the part of the EU required the existence, in Brussels, of centralized decision-shaping agencies. These were not the reasons dreamed up in Brussels by starry-eyed Europhiles. They reflect the collective wisdom of the EU's twenty-seven member states, which unanimously authorized the new institutional arrangements. (Howorth 2011, 304–5)

Ideally, the Lisbon Treaty was supposed to lay out precise terms of reference for the HR/VP, yet significant uncertainty remained about the sharing of competences as regards external relations. Part of the confusion was the uncertainty about what the HR/VP was supposed to do (most likely also added to by the decision to water down the title from 'Union Minister for Foreign Affairs', which was the initial suggestion in the constitutional convention, to 'High Representative'). On this point the Treaty remains vague as it states that the HR/VP shall

> represent the Union for matters relating to the common foreign and security policy. He [*sic*] shall conduct political dialogue with third parties on the Union's behalf and shall express the Union's position in international organisations and at international conferences. (Article 13a; see EC 2007b, 27)

At the same time, however, the Treaty states that the President of the European Council shall

> ensure the external representation of the Union on issues concerning its common foreign and security policy, without prejudice to the powers of the High Representative of the Union for Foreign Affairs and Security Policy. (Article 9b; see EC 2007b, 18)

and that the European Commission shall

> ensure the Union's external representation [...] with the exception of the common foreign and security policy, and other cases provided for in the Treaties. (Article 9d; see EC 2007b, 19)

As these conflicting passages in the Lisbon Treaty illustrate, the competences on foreign policy, let alone external relations, are far from clearly defined and leave plenty of room for manoeuvring and turf wars between the individual institutions. According to Howorth (2011, 305), it was clear from the beginning that Lisbon would not be the end, but the start of a 'process of political horse-trading and institutional jostling in which the job profiles, the procedures, the mechanisms, and the personnel of these new positions would be the prey of a host of individuals and agencies'. The European Parliament (EP), for instance, argued that 'while the EEAS should be a *sui generis* service from an organisational and budgetary point of view, it should be incorporated into the Commission's administrative structure' (Cameron 2011, 242). Clearly the EP's argument was also due to its interest in gaining influence over the EEAS through its budgetary power; however, the proposed institutional arrangement would have ensured substantial independence of the service while still facilitating its administrative establishment and organisation through its inclusion in existing Commission structures. Mainly due to pressure from member states who feared that the Commission might dominate the EEAS, it was eventually established as a completely *sui generis* service outside the Commission administration. In practice, this administrative separation is a cause of severe problems in the internal operationalisation of the service (see Chapter 5).

Beyond practical problems in the operationalisation of the EEAS, the decision to keep it completely separate was indicative of wider resentments many member states harboured against the new service – and the feared loss of competence over foreign policy matters. This also applied to the member state's decision to appoint Catherine Ashton as the HR/VP. 'The manner in which Baroness Ashton was appointed', Fraser Cameron (2011, 239–40) argues, 'reveals much about the ambitions of the member states. The appointment was never a question of foreign policy expertise (there were numerous better qualified candidates) but rather part of a balancing act in which different political groups had to be satisfied'. A member state diplomat in Nairobi echoed this assessment, arguing that 'Ashton was a deliberate pick as a very weak candidate. If she was more active, she would be led to the slaughterhouse by those who have other political interests' (Author interview, January 2012). Also for Ashton's successor, Federica Mogherini, similar observations have been made, criticising the filling of top jobs not based on experience and competence, but on quota and 'horse-trading' (Gros 2014). For Asthon, Howorth recounts how her appointment resulted in large part from a management crisis conducted by British Prime Minister Gordon Brown and his previous failure to secure any of the top EU posts for Britain:

> On the very eve of the 'appointments summit' in November 2009, it seemed that the UK would leave the meeting with empty hands – a political disaster for the beleaguered Brown. In the event, it was Jose Manuel Barroso who suggested Ashton. Did she not meet all the criteria – a left-of-centre female from a large Northern state? The fact that she was totally unknown in Europe (and virtually unknown in Britain), that she had zero foreign policy experience, and that she had never been elected to anything was less important in this process than the fact that Brown could claim a minor triumph and that Barroso would wind up with a colleague as Vice President of the Commission who, to put it diplomatically, would be unlikely to cause him sleepless nights. Thus was the appointment sealed – and nobody was more surprised than Catherine Ashton herself. (Howorth 2011, 308)

Ashton started her tenure on 1 December 2009, a year before the official launch of the EEAS. Her first months in office were therefore far from easy. Not only was she relatively inexperienced in both foreign policy and the *modi operandi* of the EU-system, but also formally the tasks ascribed in the Lisbon Treaty (Articles 9e and 13a; see EC 2007b, 21, 27) to the double-hatted position of HR/VP were only loosely defined and excessive in their scope. These tasks included:

- Conducting the Union's foreign policy
- Making proposals for the development of that policy
- Presiding over the Foreign Affairs Council
- Being one of the Vice-Presidents of the Commission
- Ensuring consistency in the Union's external action
- Ensuring implementation of decisions adopted by the European Council and the Council
- Representing the Union in matters relating to common foreign and security policy
- Expressing the Union's position in international organisations and at international conferences
- Conducting dialogue with third parties on the Union's behalf
- Leading the European External Action Service and cooperating with the diplomatic services of the member states

Moreover, in addition to these tasks ascribed permanently to the HR/VP, the most pressing endeavour for the newly created position was to design and establish the EEAS – in terms of both its administrative operationality and its thematic-diplomatic orientations. An EU official in Brussels recounted how

> [t]he first thing we did was to try to develop some of the structures; very crude structures. We started to think how the organigram could look like. Then senior post were determined, the corporate board. Then they started to build it up and recruit more staff. The requirement is that staff comes roughly to 1/3 from the Commission, 1/3 from the Council, and 1/3 from the member states. But those for Commission and Council are not required, only for the member states. At the moment we are below in headquarters, but at about 40% in the delegations. (Author interview, July 2012)

The process of establishing the EEAS included jostling competences and policy orientations between the Commission and the Council in Brussels and also between Commission delegations and member state embassies in third countries. At the same time, the EEAS was supposed to develop an identity and administrative structure as well as a niche for its operations and competences between a powerful Commission and the member states. Unlike a national diplomatic service, the EEAS is a 'new composite actor' (Berger 2012, 53) that has no homogenous Foreign Ministry – with one clearly identifiable principle – whose orders are to be followed or policies implemented. The HR/VP heads the EEAS, but she sits between different chairs and has to answer equally to the Commission, the Council and the European Parliament. The already difficult task of a national foreign minister (or diplomatic service) is therefore exponentially more complex in the case of the EEAS due to the difficulty of a negotiation process that necessitates first finding a common position between Commission and Council before representing it externally. These challenges inscribed into the position of the HR/VP by design are difficult enough to master even for a highly experienced diplomat; for a diplomatic novice they appear insurmountable. It was inevitable that Ashton would have a tough start and cause a fair amount of frustration amongst her own staff and member state officials who were supposed to cooperate with her and align their policies with the EEAS. An EU official in Brussels, for instance, recounted that

> Ashton gets a lot of criticism internally. Westcott has a good connection to her, not only because he is also a Brit, but this certainly is a reason. At the beginning Ashton mentioned to her staff that she hopes that they like the new building because she spend a lot of time choosing the colour of the carpet. I do not think she was joking!
>
> [...]
>
> She always moves around with a big entourage – like a king. (Author interview, July 2012)

With respect to the prospects of a renewal of Ashton's term in office another EU official frankly responded: 'I hope not!' (Author interview, July 2012). Also amongst national diplomats in Nairobi, Ashton was viewed very critically. Asked about the role of the HR/VP, a member state diplomat in Nairobi asked ironically: 'Who is Ashton???' (Author interview, March 2012) whilst another national diplomat clearly stated that for the sake of a more active global role for the EU '[t]he first and most important thing is: Ashton has to go! Then we can talk again' (Author interview, March 2012). Part of the frustration with Ashton was probably due to her specific inexperience in matters of foreign policy. However, the hiring procedure and a certain amount of jealously clearly also played a part as it was difficult to explain why a politically and diplomatically inexperienced person like Catherine Ashton was chosen for the top job of an emerging EU diplomacy when much more qualified candidates were excluded.

The EEAS' Early Days

The official launch of the EEAS was on 1 December 2010 – 'with ultra-discretion and no fanfare' (Howorth 2011, 316). Keeping a low-key for this landmark step in the evolution of the EU's external relations institutions came as a reaction to the harsh criticism of the HR/VP following her instalment in office and of resulting low expectations regarding the service. However, for the first time, the integrating European polity had a collective external service in the political realm – mandated by Commission and Council. Ashton had one year to prepare to make 'her' new service operational but an EU official in Nairobi recounted how this change had affected (some) delegations:

> On the first of December 2009, the Lisbon Treaty officially came into effect. Just before Christmas 2009 the information was sent out that from the first of January 2010, the Commission delegations will be transformed into EU delegations and that the External Action Service would take effect in 46 delegations – of which 26 were in Africa. So, they had about one week during the Christmas period to prepare for all the changes. Abroad, the EEAS started about 11 month before it started in Brussels. On the first of December 2010 it was officially launched in Brussels. (Author interview, March 2012)

This sudden change posed significant problems, in particular in delegations and member states' embassies around the world that were caught off-guard by the earlier-than-expected reshuffle. The upgrading of the delegations from the Commission to EU delegations was accompanied by a range of challenges such as agenda-setting, organisation, hosting and chairing of the regular meetings of Heads of Mission, Political Counsellors, Trade Counsellors, in addition to providing the necessary resources for personnel and premises. The change came equally unexpected for Spain, the country that had been preparing for the presidency to commence in January 2010. The same EU official continued:

> Of course this caused severe problems with the Spanish who had been preparing for the presidency for several months. Shortly before it started it was pretty much taken away from them. However, in light of this it was remarkable how well it worked out. A lot was negotiated on personal levels and practical arrangements were made to keep things going. (Author interview, March 2012)

A member state diplomat in Nairobi judged the introduction of the EEAS much more critically, recounting how

> [w]hen the EEAS was established, the Delegation was very surprised as it came at very short notice. Delegations often were not ready and a lot of the work was still done by the member states. Another problem was that the staff was just very inexperienced and too junior, sometimes people came straight from the university and were supposed to fulfil leading functions. There was too little capacity and experience. The member states' experiences were pushed aside. All was supposed to be established very quickly overnight. (Author interview, November 2012)

The key challenge in particular in the early days of the EEAS was to upgrade the 'coordination reflex' to a (exclusively?) joint articulation of collective foreign policy. Intergovernmental mechanisms to coordinate on matters of foreign policy had existed since the 1970s, first through the EPC then through the CFSP. The CFSP also continued to exist after the Lisbon Treaty, but it moved beyond its former almost exclusive intergovernmental character to a more binding and institutionalised articulation of foreign policy. This is not to say that the foreign policy has been transformed from an intergovernmental to a supranational policy field – it remains a national competence and the CFSP is in essence intergovernmental – but it has now been given a joint institutional structure representing all EU organs, which, next to the Commission, means first and foremost the Council and through it the member states. A key change was therefore that both Commission and member states become less credible when deviating from the joint position – as they were supposed to have agreed on a joint position through the new institutionalised processes of the EEAS. In a way, the 'coordination reflex' was upgraded to a 'coordination obligation' creating increasing pressure to align positions. Schimmelfennig and Thomas (2008, 1) describe this as *entrapment*, 'the process by which member states find it difficult to escape the dictates of substantive EU norms and thus accept (however reluctantly) norm-consistent policies that diverge from their actual preferences'. The establishment of the EEAS has further institutionalised such entrapment by making it more 'costly' to 'escape the dictate' of the collective position (Carta 2012, 23). Instead a system of cooperative bargaining has emerged 'by which veto threats are sidelined by the EU's procedural norms in favour of consultation and consensus, leading member states to adopt common policies based on mutual compromise' (Schimmelfennig and Thomas 2008, 1).

Two-step Diplomacy

These processes also hint at one of the most fundamental challenges for EU diplomacy. In contrary to a national diplomatic service, it is a two-step process. This includes, first, having to find a common internal position and, second, articulating, representing and implementing this position to the external partner(s). Necessarily these two steps involve power struggles on different levels and between different agencies, i.e. between the Commission, Council, EEAS, and member states, over how to position and articulate the collective approach. Putting the service in practice requires multiple compromises; first an internal one between *all* the member states and the Commission – often described using terms such as harmonisation, collectivisation, and to some extent also Europeanisation. Between the member states this started to emerge in a non-binding way in the 1970s in the form of the EPC; yet particularly since the Lisbon Treaty this process has increasingly also included the Commission. Second, it requires a compromise with the external partner. This second step is what national diplomacy is traditionally concerned with – the processes of negotiating (foreign political) positions and relations between sovereign states. However, for EU diplomacy, this is only the second step of conducting external relations. The first, and possibly more difficult one, is finding common positions to be projected externally. Carta (2012, 22) refers to such two-step diplomacy as '*metadiplomatic* in purpose: that is, its final goal is the acquisition of a common foreign policy to be negotiated, eventually, with external bilateral or multilateral partners', which induces to diplomats the difficulty of engaging in a 'continuous form of mediation *inside* the entity they represent and *outside*, on behalf of it' (Carta 2012, 29).

The EEAS is a novel institution that bridges these two functions. As such it is structurally different from traditional diplomatic actors. A national diplomatic service has one clear principal – the national government in the form of the foreign ministry whereas the EEAS has 'multiple centres of power and decentralised processes' (Berger 2012, 28). Contrary to a national diplomatic service the EEAS thus has several principals, two at best (the Commission and the Council), 29 at worst (Commission and each member state). Citing a senior commission official, Kuus (2014, 71) remarks how, in Brussels 'any national background is too narrow […] in national foreign ministries, there is one chain of command. Here, [in Brussels'] there are several. You have to pay attention to all of these things'. The EEAS is tasked with representing the joint position of these principals and chains of command, but the difficulty of developing such joint positions is often underestimated. Carta (2012, 85) argues that the EU continues to be 'represented by a constellation of diplomatic actors, rather than a single diplomatic star. At time the interests and preferences of these diplomatic actors converge, and at times, they conflict'. In addition and also in contrast to a national service, a difficult hurdle for the EEAS to overcome is the need to win trust and justify its existence to both its internal composites (in particular the member states) and its external partners. As regards the member states, the need for and legitimacy of a collective diplomatic service are not always self-evident. Some of the member state diplomats in Nairobi, for instance, expressed their dissatisfaction with the EEAS, often mixed with jealousy regarding the benefits available to EEAS and Commission staff in third countries:

> The EU spends so much money in Kenya, yet it is hardly visible. But they do not do anything about it. They have so much staff (150). What do they do? Why are they not more visible? They have a lot of stuff not doing anything, they all have their little agendas but do not do the things that would be really helpful for us. They are all paid very well and have very good security, also get housing help here in Nairobi. Why is that not done for us? It's our money that pays them!
>
> […]
>
> The EEAS is useless, because it does not do practical stuff and does not give us the assistance we could need. (Author interview with member state diplomat, March 2012)
> What does the EU do here? There is no political person making statements, no communications person? What do they with the money? There are more than 100 people work at the Delegation, what do they do? The member states are forced to reduce their administrations, they are being bogged down by reality of cost saving, reducing public services, at the same time the EU is apparently endlessly expanding its bureaucracies and administrations with no results. It seems to have endless resources, the countries don't. (Author interview with member state diplomat, March 2012)
>
> The Delegation has more than 150 staff and they still complain that they have too little staff to organise the meetings and so. It's a very inefficient use of money. (Author interview with member state diplomat, November 2012)

These quotes from member state diplomats illustrate some of the resentment they harbour that accompanied the establishment of the EEAS in third countries. Of course, not all diplomats viewed the EEAS negatively, yet the early phase of the EEAS was comprised of numerous struggles between the different European partners over the positioning of collective diplomacy and over the interaction with the Kenyan side (see Chapters 5–7). Also, the need for the new institution to justify its own existence within several of the member states' diplomatic services becomes clear from the quotes above. The justification of its existence was, however, not only an issue with respect to the member states, but even more so amongst the external partners, such as the Kenyan government and non-governmental and international organisations (see Chapter 6 for a more detailed account of how the EEAS is viewed by its external partners). The key problem was that the function of the EEAS was not immediately clear to its partners. The Commission had been a long established partner in issues of development and trade, and the member states had been long established actors with regards to diplomacy and foreign policy. The role of a new diplomatic, foreign political service of a collective Europe remains unclear – as does the separation between European Commission, European Union, Common Foreign and Security Policy, and European External Action Service. As the new, and only vaguely defined, institution in this complex array of European external relations instruments, the EEAS had to fight for recognition and legitimacy. The two-step diplomacy process is certainly hindering this because the lengthy processes involved in the first step, finding a common position amongst the European partners, prohibits the flexibility of the EEAS in the second step, its interaction with its external partners – the *actual* diplomacy. This lack of flexibility, and sovereignty (because of the need for continuous feedback with its multiple principals) on the part of the EEAS hinders the upgrade of its political role amongst the cooperation partners. Nevertheless, the service's multiple principals, in particular the Commission's inclusion in articulating collective foreign policy, counteracts 'strict intergovernmentalism and favour[s] institutionalised bodies' (Berger 2012, 28). *If* there is a consensus between the different principles, the EEAS will therefore contribute significantly 'to ensure coherence, since it possesses the resources necessary to control and monitor the planning as well as the implementation process [yet] due to the unchanged decision-making procedure in which unanimity remains the ultimate rule, progress is prone to be a victim of political will' (ibid., 83).

The EEAS in the EU's wider External Relations Portfolio

The term 'external relations' is used in this book to describe the entirety of the EU's interaction with external partners. With respect to Africa-EU relations, the main policy fields are trade and economic policy, development policy, security policy and foreign policy. The level of integration and collectivity in each policy field, of course, varies considerably with the most integrated being trade and economic policy as they are almost exclusively supranational due to the Commission's mandate to handle these policy fields on behalf of the member states.

Economic relations with most African countries, including Kenya, are currently organised under the scheme of the Economic Partnership Agreements (EPAs) between the EU and the African, Caribbean and Pacific (ACP) regions. Officially, the EPAs aim at 'promoting trade between the two groupings – and through trade development, sustainable growth and poverty reduction' and 'set out to help ACP countries integrate into the world economy and share in the opportunities offered by globalisation' (EC 2013a). They have, however, also received fierce criticism as imposing terms of trade and conditionalities favourable to Europe's advanced economies on the EU's relations with African countries (Holden 2009; Bachmann 2011a; Hurt 2012).

Development policy is a hybrid policy field, comprising of long-established supranational components in the form of European Development Funds (EDF), ad-hoc intergovernmental cooperation between member states as well as bilateral cooperation between individual member states and the host country (Carbone 2007). The single biggest instruments of the collective development policy towards African countries are the EDFs, the first of which was launched in 1959. The most recent one, the 11th EDF, covers the time period of 2014–2020 and provides the funding for the EU supranational development policy (EC 2013b; Tannous 2013; EC 2014b).

Also security policy is, perhaps surprisingly, ambivalent with respect to African-EU relations. Traditionally considered the 'last bastion' of national policy-making with at best some intergovernmental coordination, security cooperation with African countries is often different. Clearly, individual member states continue to have national security interests and long-established military cooperation with African countries, in particular France and the UK. However, in times of tight budgets in many European countries, European societies increasingly display a certain fatigue with respect to national military missions abroad. Also a more multilateral engagement is usually perceived as of greater legitimacy than unilateral action – in particular when undertaken by a former colonial power. As a result, European security policy in Africa is surprisingly collective, including direct European missions, such as the missions to the DRC, South Sudan and the Horn of Africa, as well as the financing of African Union missions carried out by African troops, such as the African Union Mission to Somalia (Vines 2010; EEAS 2013b; Germond 2013).

Foreign Policy, in many ways, is the least supranational of these policy fields. Whilst security policy has become increasingly collectivised – partly motivated by countries' fatigue at 'getting their hands dirty' – foreign policy has remained firmly in the hands of member states. Clearly the 'coordination reflex' described above that emerged with the EPC in the 1970s has subsequently intensified, not least to a more institutionalised Common Foreign and Security Policy. However, the final say over diplomatic relations and foreign policy towards African countries has always remained with the member states. Before the EEAS, no institution existed that was mandated to speak politically on behalf of the European Union. This has changed as a result of the Lisbon Treaty and with the launch of the EEAS (see above). The metadiplomatic two-step process described above of course presents a tremendous difficulty in achieving a nascent European diplomacy. Yet once the first step is completed, i.e. a common position found, the EEAS, mandated by both Council and Commission, has the authority to articulate and conduct diplomacy and foreign policy on behalf of the European Union (including all its member states). Still foreign policy is not completely supranationalised. Much like development policy, member states continue to coordinate intergovernmentally through the CFSP, as well as to conduct bilateral foreign policy and diplomatic relations with third countries. The EEAS is an *additional* institutionalised attempt at harmonising intergovernmental coordination and bilateral policy conduct, but does not replace these. According to Berger (2012, 15) it is 'indeed a capable actor, but in terms of external coherence it is nonetheless vulnerable to the political will of the individual member states'. To some extent collective foreign policy thus remains intergovernmental and national/bilateral, yet, through the EEAS it has gained a supranational component (Carta 2012; EEAS 2013a).

Looking at these four key policy fields reflects a more general trend of levels of integration in different policy fields. It is easy to observe how internal levels of integration are generally reflected in the Union's external relations. Mostly, further steps of integration initially take an economic route, followed by a more political aspect. Just as within the EU, also with respect to its external relations, economic integration is more advanced than political integration. Competence in external trade policy and economic relations rests with the Commission. Diplomacy and (foreign) political relations received an impetus towards integration through the Lisbon Treaty, yet they remain fragmented and subject to individual member states' preferences. In a pre-Lisbon context, I argued elsewhere that 'whilst Europe has long succeeded in acting with relative unity on economic issues, the interests of individual member states/governments often preclude a more active collective geopolitical role. This political fragmentation is a feature of the uniqueness of *and* limits to the European integration process' (Bachmann and Sidaway 2009a). This observation still largely holds true. With the EU's geopolitical strength mainly rooted in its economic power, its collective role in the world is more pronounced in the economic than in the political arena (see Bachmann 2013a).

At the same time, it can be observed that the experience of integration in development policy traditionally set the path for external integration in other policy fields, including for instance diplomacy or foreign policy (see Figure 1.1). Initial steps towards integration in external relations were provided for in the Treaty of Rome with respect to collective development policy in the form of the supranationally administered European Development Fund. Development policy was thereby a less contentious and prestigious policy field of external relations than foreign policy – often seen as a symbol of national pride and independence. Collectivisation required less commitment and compromise than ceding sovereignty in the prestigious fields of diplomacy and foreign policy – yet it allowed a test of coordination mechanisms and collectivisation abroad to take place. These experiences – both positive and negative – influenced the processes of collectivising foreign policy – first in the form of the EPC, then of the CFSP and most recently in the EEAS. Over more than half a century of collective development policy since the Treaty of Rome, a multi-layered system of European development policy has emerged, involving supranational, intergovernmental and bilateral components and making the EU the largest donor of official development assistance in the world. Collective European diplomacy and foreign policy is far from having such a status in the international system. Indeed, it is highly unlikely that foreign policy will become as collectivised as trade policy as sovereignty largely remains with the member states and European foreign policy will retain its bilateral and intergovernmental components. However, the EU's institutional restructuring through the Lisbon Treaty envisioned a significant upgrade of the EU's collective role in the world – including with regards to diplomacy and foreign policy. The EEAS, designed in the wider EUphoric geopolitical context of the mid-2000s, is thereby the key institution for enacting the EU's global role.

Chapter 2

European Geopolitical Discourse and the EU as a 'Model' in Times of Crisis

The previous chapter has traced the evolution of European integration, in particular with respect to the development of instruments for the collective conduct of external relations, eventually leading to the creation of the EEAS. In this chapter, I will trace how the development of these instruments has been accompanied and shaped by related geopolitical discourses, creating visions and ambitions, and justifying legitimacy for the EU as a global actor. The analysis presented in this book is situated within the wider school of thought of critical geopolitics. It seeks to capture how geopolitics is historically and spatially conditioned, how it is framed by wider political and economic constellations, how it is embedded in local and wider international contexts, and how it is constituted both materially and representationally (see Introduction). Within this broader understanding of geopolitics, this book emphasises the social construction of geopolitics. I study perceptions and interactions, the practices and processes of how collective European diplomacy is enacted by key individuals of African-European relations in Nairobi. In terms of the tripartite differentiation of critical geopolitics into formal, practical and popular spheres (see Introduction), the focus of this book is on the *practical* sphere of how collective EU agency is enacted and perceived.

This chapter, however, lays the *formal* geopolitical foundation for the research by providing the wider political and economic context and tracing the evolution of how the EU is positioned as a global and development actor in key discourses on the EU's role in the world. It outlines how European integration has been accompanied and influenced by civilian and normative power debates, whereby the integrating Europe has often been idealised as a geopolitical model for regional integration to be emulated elsewhere. In addition, it also illustrates how external cooperation partners indeed perceive(d) of the EU as such a model and expect(ed) it to assist African regional organisations in their integration efforts. Such expectations and the perceived appeal of the 'European model' are thereby a key factor for legitimising a global role for the EU. Narrative and reality of integration thus went hand in hand as the *formal* and *practical* sphere of the geopolitics of European integration (and external positioning). The emergence of the Eurocrisis, however, has brought to the fore serious doubts about the functioning of Europe's model – roughly around the same time the Lisbon Treaty and the EEAS sought to establish the more coherent and influential global role of the EU (Holden 2012; Jones 2012; Murphy 2013).

European Integration and its Narratives

The current system of political-economic organisation across large parts of the European continent has been significantly shaped by a process of regional integration. This process, of course, predates the European Union. After centuries of describing certain commonalities of Europe as a more or less homogenous cultural region (Davies 1997; Grataloup 2009), World War One fuelled nationalism and the defining of identities in terms of national categories. After the destruction of the continent through the Napoleonic wars at the beginning of the nineteenth century, Europe witnessed a phase of relative internal peace. Increasingly confident and aspiring nation-states focused on imperial expansion beyond Europe with Germany's industrialisation and increasing geopolitical ambition during the second half of the nineteenth century changing the balance on the continent. The Austro-Prussian war of 1866 and especially the Franco-Prussian war of 1870/71 were first indicators of a growing German/Prussian unease about its geopolitical situation on the Continent and its willingness to aggrandise at any cost (Hugill 2008). It thus also heralded the return of large-scale war to the European Continent. Less than half a century later, the industrialised warfare of World War I revealed the new dimension of destruction that technological advancement during the previous hundred years had brought to military conflict. With its victorious intervention into the war, the United States also cemented its new global lead role – replacing Britain as the world's dominant power. The United States pursued a different geoeconomic strategy. It gained economic strength yet was politically hesitant at playing a global hegemonic role. Within this phase of geopolitical transition, the British, broadly liberal, geoeconomic doctrine was slowly but increasingly replaced by a more protectionist geoeconomic orientation favoured by a geopolitically maturing United States (Meinig 2004; Hugill 2009).

With economic protectionism becoming more fashionable, nationalism, too, was on the rise. Before World War I, 'the nation-state – tariff-protected, increasingly bureaucratized, and highly nationalistic' had already emerged 'as the dominant geopolitical actor of the late 1800s' (Hugill and Bachmann 2005, 161). The war thus furthered an already existing climate of increased identity-building and the more central role of the nation-state as the predominant unit of political identification and organisation. At the same time contemporary observers in the 1920s were aware of the danger of another prolonged phase of warfare on the European continent. It is in such a climate that the Paneuropean Union was founded by the Austrian Richard Nikolaus Coudenhove-Kalergi in 1923 and quickly grew to include prominent members[1] from almost all European countries open to the idea of a sustainable and deep integration of Europe's different peoples. During the economic crisis of the late 1920s, the Pan-European Union laid out integration visions for Europe to pursue during the times of the then crisis, something that was

1 Such as Albert Einstein, Thomas Mann, Franz Werfel, Salvador de Madariaga, Charles de Gaulle, Aristide Briand.

seen as the only way to avoid another war on the continent (Coudenhove-Kalergi 1934). Their plans failed. Yet, after World War II, these early ideas of European integration were revived by the architects of the Treaty of Rome before the 1960s saw an intensification of integration measures (see Chapter 1).

A European Geopolitical Discourse

In the early 1970s, François Duchêne's notion of Europe as a civilian power laid the foundation for a wider engagement with the geopolitical identity and role, and the future direction to take, of an integrating Europe. It was based on the experiences of two world wars in the first half of the twentieth century and emerged at a time of faltering American hegemony during the Cold War. Duchêne advocated the 'domestication' of relations between states, i.e. that international relations should be dealt with in structures and modes of domestic policy conduct (Duchêne 1973). According to Duchêne, (Western) Europe should set an example for taming anarchy in the international system and providing a 'civilianized' alternative to the first half of the twentieth century and the realist world order of the Cold War (Bachmann and Sidaway 2009b; Manners 2010). Politically, the subsequent discourses prioritised multilateralism, supranational integration, institution-building, the rule of law and the restriction of the use of force in international politics. However, they also had strong normative orientations, focusing on human rights, democracy promotion, social and economic fairness, etc. (Maull 1993; Kirste and Maull 1996; Maull 2005).

In contrast to these, often idealistic, debates, the Australian international relations scholar Hedley Bull took a very sceptical position towards civilian power ideas and any form of power that is not backed by effectively deployable military force. He argued that all power and influence of 'the European Community and other such civilian actors was conditional upon a strategic environment provided by the military power of states, which they did not control'. Consequently Bull disagreed with Duchêne on the point that the EC is a significant actor in international affairs or might develop into one (Bull 1982, 151). He emphasised that 'there is no supranational community in Western Europe but only a group of nation states' and viewed the EC as an instrument of cooperation among governments with a history of 'endemic mutual conflict'. Bull was also sceptical about the devaluation of military power in favour of an increasing reliance on cooperation and negotiation. In fact he saw military power as an underlying necessity without which negotiations could not even take place. According to Bull (1982, 156) there is

> no contradiction between reliance on détente, on the pursuit of political settlements, the promotion of trade and understanding, and the reliance on a balance of power: it is the security afforded to each side by effective defences that provides the confidence required to work for change.

For Bull, the countries of Western Europe could no longer rely on the American provision of security throughout the 1980s and therefore should 'take steps towards providing themselves with nuclear deterrent forces that will in due course take over from the United States the function of neutralising any Soviet nuclear threat' (ibid., 157). Bull thus advocated the creation of a nuclear-equipped European military alliance as the most important step of integration, with other forms of European integration such as the European Economic Community serving this military alliance.

Twenty years later, Ian Manners (2002) developed the notion of Normative Power Europe (NPE) as a response to Bull's realism and the geopolitical events of the 1990s. Manners' NPE approach has since become the key reference point for EU geopolitical discourse – often heavily contested. In many of these debates, the question of the EU's international identity remains a key challenge as it is more difficult to grasp than the question of what role the EU plays in which policy area. Forsberg (2011), for instance, argues that EU foreign policy analysis frequently invokes the NPE notion, yet fails to systematically define what normative power actually means. Forsberg further suggests a distinction between the EU's normative orientation and its power to implement policies. Instead of a yes/no definition of the EU as a normative power, he advocates conceiving of 'normative power' as an ideal typical notion, with which the EU (or other actors) complies in certain situations and not in others. Similarly, Eriksen (2006) attempts to define certain criteria for normative power in order to position the EU as a potential cosmopolitan polity (see also Beck and Grande 2004), thereby raising questions of the EU's *normative* orientation in foreign policy (see also Zielonka 2006; Zielonka 2008).

Other, in particular realist accounts, instead question the *power* aspect of NPE, with Risse (2012) expressing concern with how the EU's internally constructed identity as a civilian power shapes its foreign policy orientation. He points to a lack of *power* of the EU as a geopolitical actor, i.e. the willingness and means to implement (normative) foreign policy objectives through collective European institutions (see also Maull 2005). Similarly, Biscop and Coolsaet (2003, 30) argue that 'without the willingness to apply pressure, sanction and, if need be, force, EU external action will not acquire the credibility it needs to be effective'. Referring to Solana, Menon identifies a general sense of European values that includes a 'shared responsibility to protect populations from genocide, war crimes, ethnic cleansing and crimes against humanity'. He argues that the EU's inability to do so is 'hardly acceptable, particularly given the proliferation of ambitious rhetorical claims concerning the EU's role as a positive force in world politics' (Menon 2009, 245). Military capability and a further collectivisation of its international role, to Menon, are therefore essential for the EU to uphold its civilian and ethical ideals – and by consequence its credibility as a global actor (Menon 2014). Based on the lack thereof in certain situations, Maull (2005) conceives of the EU not as a civilian power, but rather as a civilian force in the international system. For realists, like Hyde-Price (2006, 220), the EU is therefore 'not a sovereign actor in its own right, but acts as a vehicle for the collective interests of its member states, primarily for shaping an international "milieu" or advancing 'second-order concerns' of its member states, such as ethics (ibid., 222–3). More popularly, the neoconservative Robert Kagan (2003, 3) denies a collective European global role entirely. He argues that 'Europe is turning away from power, or to put it a little differently, it is moving beyond power into a self-contained world of laws and rules and transnational negotiation and cooperation' (Kagan 2003, 3), partly by choice, however, also because of a lack of another alternative in light of the military mismatch with the United States.

Other facets of the geopolitical debate on the EU include, of course, engagements with the causes and effects of the crisis of the Euro. Murphy (2013), for instance, explores EU identity construction, focusing on its link with European Monetary Union (EMU). Tracing the evolution towards EMU, Murphy argues that EMU has not fulfilled its initial objective and is, in light of the current financial crisis, instead stimulating negative perceptions of collectiveness in large parts of the European population (see also Jones 2012). Other debates include suggestions for conceiving of the EU as a 'demoicracy' – a polity of multiple *demoi* that can serve as a normative benchmark for transnational non-domination and transnational mutual recognition (Cheneval and Schimmelfennig 2013; Nicolaïdis 2013). Manners offers yet another notion of the EU, that of a 'European Communion' centred around shared relationships and the ambitions of 'achieving reconciliation and equality in order to overcome historical experience; recognizing and respecting identity and cultural diversity; and acknowledging an ethic of cooperation rather than coercion' (Manners 2013, 483). Many more descriptions of the EU, for instance as model power, ideal power, empire, etc. exist (Manners 2010; Cebeci 2012; Kaldor 2012; Wong 2012; Telò and Ponjaert 2013), yet the normative and civilian power debates remain key reference points for the wide body of literature on the EU's geopolitical identity and role.

These debates contributed significantly to shaping a discursive and intellectual environment within which the social construction of the EU's international agency emerged and evolved (Bachmann and Sidaway 2009b). As such they are part of the broader sphere of formal geopolitics and the 'production of global political space by dominant intellectuals, institutions, and practitioners of statecraft' (Ó Tuathail 1996, 185).

Geopolitical Ambitions and Deducted Legitimacy

Whilst the liberal-ideational discourses of civilian and normative power have always been contested by realist accounts (see above), they play a key role for articulating wider geopolitical ambitions and visions of an integrated Europe, and for legitimising its global role. In their comprehensive account *The European Union as a Global Actor,* Bretherton and Vogler (2006, 56–7) identify three complementary roles for the EU in world politics that have emerged in the second half of the twentieth century: the EU as a model for regional integration, the EU as a promoter of normative and civilian values, and the EU as a geopolitical counterweight to the USA. The lead discourses, civilian and normative power, are inextricably linked with these roles. They have always had both a descriptive and prescriptive character in that they aimed to describe and analyse the process of European integration and simultaneously laid out visions for future integration and geopolitical roles to attain (Nicolaïdis and Howse 2002, 770). This prescriptive component traditionally included ambitions towards a more influential, collective, global role. Notions such as the 'collective exploitation of resources' can be found in the writings of the Paneuropean Union of the 1920s (Coudenhove-Kalergi 1931), in the early integration documents of the 1950s and 1960s (see Chapter 1), as well as in the integration narratives starting with Duchêne's civilian power (see Bachmann and Sidaway 2009b; Bachmann 2011a). Such notions of expansion and imperialism were also subject to criticism from the beginning of geopolitical integration discourse. Along with the accession of the UK, Denmark and Ireland in 1973, Johan Galtung (1973) warned of an aggrandised European Community's imperialistic undertaking. The common market, he argued, 'is considerably *more* than a "market"; it is a struggle for power, for world power for Western Europe' (17). He viewed the European community as 'a superpower embedded in the environment of other superpowers' (119), following one basic formula: 'take five broken empires, add a sixth one later, and make one big neo-colonial empire out of it all' (16).

Even though far from an outright rejection of European integration in general, Galtung saw a continuation of the general attitude during the colonial period whereby Europe viewed itself as the 'most civilised and best governed of all the world regions' (Bassin 1991, 3) and therefore ascribed itself the right to spread its model of political and economic organisation to the rest of the world. Clearly European countries have a long history of attempts of forceful projection of what were considered superior and civilisational spaces (O'Loughlin and Van der Wusten 1990; Sidaway and Power 2005). Geopolitical writings have always been part of this history and despite their heterogeneity they had 'one common theme: the production of knowledge to aid the practice of statecraft and further the power of the state' (Ó Tuathail and Agnew 1992, 192). The EU is not a state though, yet it is a polity exercising transnational statecraft within its borders and beyond. Despite Europe's imperial past (and present), I would argue that contemporary collective external relations take a different character from those of previous European imperial and colonial missions to 'civilize' the world. Nevertheless, those histories cannot easily be dismissed and a certain tendency amongst European countries, and the EU, to 'repress spectres of European empires' (Kennedy 2006, 135) remains. Indeed European imperialism is often viewed as a practice from the past, while Hooper and Kramsch (2007, 527) criticise a Europe that is

> oddly unreflexive about its own imperialisms, past and present [...] The result is a geopolitical analysis which not only precludes recognition of the

> spatiotemporal complexities of empire, but masks Europe's current complicity in the production of exploitative and oppressive relations within as well as beyond its newly minted frontiers.

In the absence of such a critical reflection on the past and present imperialism of European countries and EU institutions, the history of European integration (up to the outbreak of the financial crisis in 2008) functioned as a legitimation for the external projection of the European 'model'. From a historical perspective the initial objective of post-World War II integration in Europe – i.e. the creation of peace and prosperity – has functioned surprisingly well. Moreover, the key premises of the civilian power visions of the 1970s – i.e. rule of law, multilateralism and supranationalism, the restriction of the use of force in international politics, etc. – have largely been realised in the European Union (Bachmann and Sidaway 2009b; Bachmann 2013d). With the accession of the former autocratic regimes of Spain, Portugal and Greece in the 1980s and post-socialist countries in Central and Eastern Europe in the 2000s, the European integration process included more countries into its political-economic system of civilianised structures. It is from the (perceived) success of these processes that the EU draws the ambition for advancing its model globally, as '[t]he civilian power employed in the EU's own region-building', Hettne and Söderbaum (2005, 538) argue, 'is also being projected in its external relations as the preferred world order model'.

This missionary ambition is particularly pronounced in EU relations with supposedly smaller or weaker countries and regions. The EU's enlargement programmes and the European Neighbourhood Policy (ENP) are model examples of the projection of European structural power to impose an idealised EU system through 'positive and negative incentives', i.e. the carrot and stick method. To countries in the EU's former and current neighbourhood, the carrot constituted an eventual accession to or favourable association with the worlds' largest integrated economy (Holden 2009). To African countries, the carrot of accession does not work, yet the missionary ambition of promoting a civilianised system remains equally pronounced. Such carrots used in Africa-EU relations mostly took the form of development aid and favourable trade conditions. The narrative of successful regional integration, centred on the creation of peace and prosperity, also functioned as an attractive means of enhancing European influence in Africa – in particular to countries and regions that had suffered from political instability, unrest and physical violence.

In addition, the lead geopolitical debates – civilian and normative power – have played a crucial role in the processes of projecting Europe's 'world order model' externally. First, a certain 'missionary' ambition for advancing a civilian or normative orientation abroad has always been inherent to these discourses; they specifically include a *power* dimension that refers to the willingness and ability to promote a civilian or normative order in the international system (Kirste and Maull 1996; Bachmann 2013a). Second, they emphasise the superiority of Europe's civilian/normative geopolitical model *vis-à-vis* less desirable others by positioning the EU as a 'better' global actor (see Diez 2005) – both in historical terms and with respect to other geopolitical trajectories and actors. The civilian power grand strategy, according to James Rogers, has always sought to define a 'better' Europe against its own 'chronopolitical opponent (war, butchery, ethnic cleansing and so on)' (Rogers 2009, 842). Civilian power debates have primarily emerged and intensified at times of geopolitical flux whereby an integrating Europe has always claimed to be a 'better' alternative. These times of flux include the superpower rivalry of the 1970s, the end of the Cold War in the 1990s and the neo-conservative re-orientation of US foreign policy under George W. Bush in the 2000s (Bachmann and Sidaway 2009b). Also during the 2000s, the EU claimed the superiority of its geopolitical model *vis-à-vis* Russia, which had lost the ideological battle of the Cold War and the impoverished population of which was enduring similar shortcomings to those of many countries in the Global South. Against this was also the still limited appeal, particularly in cultural terms and issues of soft power, of China. In combination, the missionary ambition and the narrative of a better geopolitical option positioned the EU as an appealing and favourable cooperation partner and a model for African regional integration to emulate (Bachmann and Sidaway 2010; Lenz 2012; Murray and Moxon-Browne 2013). In the following, I will briefly indicate how the EU is not only positioned as a model in the discourse introduced above, but also how it was perceived in such a way and was expected to play a more active geopolitical role by key African and European development practitioners.

The External Appeal of the Regional Integration Narrative before the Crisis[2]

The results shown in Figure 2.1 are based on empirical data obtained as part of an earlier research project on EU development cooperation between 2006 and 2009 and illustrate respondents' expectations on the EU's geopolitical role, i.e. in their view what role *should* the EU be playing. They indicate three main tendencies regarding expectations of the EU's geopolitical role: 1) the EU as a model for a civilianised international system and regional integration, 2) the EU as a counterweight to the US, and 3) the EU as a partner for African countries and a manager of globalisation.

The EU was often called upon to act as a normative, value-based actor and as a model for regional integration, responses which implied that the EU should promote this model for a system of political-economic organisation to the wider international level. During an interview at the African Union headquarters in Addis Ababa an AU official denoted the EU as

> a role model for interdependencies and multilateralism, that should show others how to deal with global issues on a multilateral basis and through

2 This section is partially based on Bachmann (2013a).

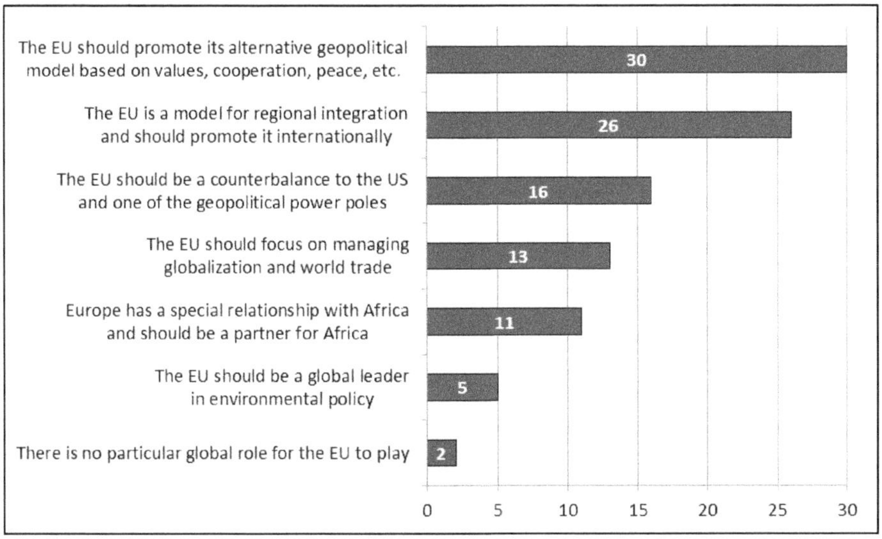

Figure 2.1 Expected geopolitical roles for the EU (2007/2008)

Note: Based on interview data with 55 African and European individuals in the field of development cooperation. Informants could give multiple responses.

Source: Adapted from Bachmann (2009, 76).

> interdependencies. It should export its peaceful foreign policy and its strong commitment to peace. There the EU has a different philosophy to the US. It should share with the world the values and achievements it has developed internally, such as human rights, democracy, technological achievements. The EU should export those values to the world. (Author interview, February 2008)

The notion of the EU as an 'exporter' of internally developed values to the international system frequently appeared in the literature and interviews. An (American) economist for the World Bank in Nairobi described the 'EU as role model for the whole world, with free movement of labour and capital' and advocated for the EU to 'transfer its interior structures to the world' (Author interview, February 2008). Similarly, another AU-official in Addis Ababa suggested that

> [i]n a globalised world the EU should share the values it has developed and realised internally with the entire world. Share with the world the tools of regional integration and act as a model. Share with the world its infrastructure masterplan, and also its principles of democracy, anti-corruption, human rights, no torture, press freedom, and all the other freedoms. Share with the world the regulation mechanisms that allow it to impose the level of civilian structures it has developed internally onto all 27 member states. Share the common values

> and the way how it managed to have those common values respected throughout the entire union. (Author interview, February 2008)

The quote touches on many aspects outlined in the literature on civilian power and Europe's role in the world; in particular it underlines two of the three roles suggested by Bretherton and Vogler (2006, 56–57). It also echoes both Duchêne's (1973, 19–21) call for a 'domestication of international relations' through the EC as a 'force for the international diffusion of democratic and civilian standards' and Maull's suggestion of a 'transfer of the interior level of civilianized structure to the international system' (Kirste and Maull 1996, 301).

Also the second role mentioned above and the final one suggested by Bretherton and Vogler, the EU as a counterweight to the USA, was repeatedly mentioned during interviews. An economist from the Kenyan Ministry of Planning and National Development, for instance, suggested that the EU 'now plays a role like the Russians during the Cold War – as a counterbalance to the US. The EU should play a stabilising role in the world and balance out the power structures. Partly it is doing so' (Author interview, April 2008). This role of the EU has indeed been cited regularly, not only during the interviews, but also traditionally in geopolitical discourse on the topic (Duchêne 1973; Bretherton and Vogler 2006). However, it is not in the focus of the argument laid out in this book as I am not primarily interested in geopolitical power struggles between Europe and the US or China, but in African-European relations and how they are perceived and enacted. But the third of the roles mentioned above – that of the EU as a partner for African countries, specifically as regards the management of globalisation – is of higher importance to the argument laid out here. It features prominently in informants' expectations of the EU's international role and it has also been suggested in geopolitical discourse. Jacoby and Meunier (2010, 299), for instance, view the management of globalisation as the 'primary driver of major European Union (EU) policies over the past 25 years'. Also, as early as 1993 and in light of the foundation of the European Union, Hill (1993, 311) suggested that the EU become the 'principal voice of the developed world in relations with the South […] and the principal interlocutor with the poor majority in the UN'. This is indeed a role that is not only frequently claimed in statements by EU personnel and official documents (see Rutazibwa 2010), but also widely expected by African informants interviewed as part of this research. A NGO coordinator in Nairobi, for instance, argued that 'the question is not if we like globalisation or not; it is a fact. The question is, how we manage it and how do we make it more humane? The EU plays a very important role in this process' (Author interview, November 2007). On the model character of the EU and the partnership with the ACP countries, an official at the East African Community (EAC) made the following point:

> The EU has gone through a process of incorporating the Central and Eastern European (CEE) countries. It therefore knows and understands the difficulties weaker countries encounter and it is a good position to assist them to be included

> in the world economy. The EU's goal in its assistance to the CEE countries has been their inclusion into Europe and into the world economy, the partnership was not one of dominance because it has been realised that a fair partnership is in the mutual interest. The idea was to create economic opportunities for the weaker partner and include them, thereby using their strength. (Author interview, March 2008)

Adapting this to European relations with the ACP-countries, another EAC official argued that 'the EU has been very supportive of the ACP-group in WTO negotiations. It is seen as a partner by ACP countries in that respect. The ACP can exercise some influence globally through the EU' (Author interview, March 2008).

These accounts are clearly debatable.[3] There is a swathe of literature that critically points at some of the highly problematic aspects of the EU's relations with developing countries (Hurt 2003; Gibb 2006; Nixson 2007; Holden 2009; Rutazibwa 2010; Hurt 2012). In the wider context of EU geopolitical discourse, I have elsewhere pointed out the civilian/power dilemma in perceptions of the EU's international role and identity, something which is based on the coexistence of largely positive perceptions of the EU's *civilian* geopolitical role model for regional integration and often negative perceptions of the EU as a geoeconomic *power* associated with past and present forms of European imperialism (Bachmann 2013a). The negative perceptions and ongoing facets of imperialism, associated in particular with EU trade policy and economic relations with developing countries, shall in no way be neglected here (see also Bachmann 2011a, 2013b), but the point to emphasise is that, due to its civilian power orientation the EU has long had a substantial appeal. Its history of regional integration was widely perceived as a success, resulting in expectations of external cooperation partners on the EU to support other countries/regions in undertaking similar development (see above). As will be shown in Chapters 7 and 8, such expectations have changed with the emergence of the Eurocrisis.

The Treaty of Lisbon and the Outbreak of the Crisis

For the EU, the decade following the turn of the millennium was a time of substantial restructuring. The Convention on the Future of Europe, established in 2001 and headed by former French President Valéry Giscard d'Estaing, was charged with drafting a constitutional treaty for a soon-to-be enlarged Union. It was a time of general euphoria for the perceived success of the integration process – well before the economic downturn which led to the outbreak of the crisis. The Maastricht Treaty had been signed less than ten years previously; Austria, Finland and Sweden acceded to the Union in 1995; and the Euro had

---

3 For alternative views on the EU's East Enlargement, see Henderson (1999), Agnew (2001), Moravcsik and Vachudova (2003).

just been introduced as the common currency without major difficulties. In 2009, the suggestions of the Convention were partially realised in the form of the Lisbon Treaty. In the meantime the EU had witnessed the accession of 12 Eastern and Southern European countries (plus Croatia in 2013). The Lisbon Treaty, then, was a key moment in the EU's institutional evolution and a crucial step for consolidating the path of European integration. With it, the European External Action Service (EEAS) was an equally crucial step for adapting the EU's external role to its internal level of integration and positioning it globally as a more cohesive and influential polity (see Chapter 1). Equipping the EEAS with a collective political mandate from Commission and Council marked a significant legal-institutional shift in the conduct of EU external relations and signalled the intention to provide an institutional basis for the EU to function as an influential global political actor on a par with powers such as the United States and China. Both the Lisbon Treaty and the EEAS thus responded to the growing international ambition of the 2000s, an ambition fuelled by the perceived success of the EU's path of regional integration, the decreasing international reputation of the United States during the Bush presidency, the aspiration to become one of the power poles in a new multipolar world, and the respective external expectations of playing a more active global role.

Key to this growing international ambition, as well as to the deducted legitimacy and the external expectations, however, was the (more or less) credible narrative of a regional integration process delivering peace and prosperity. Packed in with wider notions of civilian/normative power, this narrative has always positioned Europe as a 'better' global actor *vis-à-vis* (a) geopolitical other(s) and as an appealing, more civilian, conciliatory and cooperative, alternative within wider geopolitical power struggles – in particular during times of major geopolitical flux over the past 40 years (Bachmann and Sidaway 2009b). The current situation is different. Europe is at the heart of the problem, severely threatening the global financial architecture and the functioning of the global economy. Unlike previous times of geopolitical flux, it cannot merely react to the changes occurring around it because European problems are at the centre of that flux. As such, it cannot position itself as a 'better' geopolitical alternative because it has no viable vision or narrative to offer. The often invoked *Schicksalsgemeinschaft* (common destiny) has become more pronounced in recent years, albeit often in a negative way (Murphy 2013) leading to heightened levels of resentment and nationalism towards the 'other' Europe – often in terms of north vs. south binaries. These binaries are likely to have a very negative impact on the sense of collectiveness and European identity in the population – furthering the problem of the detachment of the population from EU institutions, integration processes and narratives, and ultimately of the legitimacy of a collective European polity (Bachmann and Belina 2012).

Ironically, it is just at the time when the EU is implementing the Lisbon Treaty and the EEAS in order to establish itself as a more powerful global actor that it is undergoing the most severe crisis since its foundation and the functioning of the 'European model' is being seriously questioned. All of the roles identified by

Bretherton and Vogler (the EU as model for regional integration; as promoter of normative and civilian values; and as counterweight to the USA) have increasingly been challenged in the current decade. In light of the crisis there are growing doubts about the functioning and benefits of regional integration per se. The EU's normative orientation has been seriously undermined by its anti-social austerity policies and increasingly restrictive migration policies (Houtum and Boedeltje 2009). While finally, due to its internal divisions and meandering, the EU appears to be losing its role as a counterweight to the USA and China (Campbell 2008; Toje 2011). All three of these roles, and the aligned civilian and normative power debates, are thereby losing their spark as the geopolitical conditions against which these debates were developed have fundamentally changed with the end of the Cold War, the emergence of the Eurocrisis and the rise of China and other emerging economies. Some of the initial ideas, such as the creation of freedom and peace in Europe and the integration of Eastern European countries, have largely been realised – and are now widely taken for granted. Other aspects, like an 'ever closer union', seem even less popular in light of growing Euroscepticism. Yet other notions, such as social balance and economic prosperity, now appear as utopic dreams far detached from the reality of a collective Europe in times of financial crisis.

These developments and Europe's internal problems of course also influence the way the EU is seen externally (see Chapters 7 and 8). Its credibility and legitimacy as a global actor is based on the narrative of functioning regional integration and it is through the perceived success of building a peaceful and relatively prosperous political entity that the EU gains geopolitical agency and is seen as a legitimate global actor whose model appears appealing for emulation elsewhere (Bachmann 2013a, 473). If this model stops working, the associated narratives lose their credibility, and the EU – as an actor whose geopolitical agency is built on such narratives – loses its legitimacy. At the same time, other global actors and development models offer viable cooperation alternatives for African countries. Most often cited in this regard is China with its state-orchestrated capitalism, offering cooperation free of conditionalities associated with human rights, social and environmental demands and thus a model of fast economic growth that is outside the traditionally hierarchical relationship with Europe.

Crisis, Discourse and Geopolitical Legitimacy

Following a critical geopolitical approach, this chapter has highlighted the centrality of the regional integration narrative, tied to civilian and normative power debates, to the EU's international role and identity. These narratives and debates, or rather the credibility of these narratives and debates, are what EU international legitimacy rests on. Whilst the effects of the European integration process (such as free movement of people and goods, the Euro, Erasmus, etc.) are now being widely taken for granted, the EU is often still not considered as an actor *per se*. Unlike a state, it continuously has to explain the reasons for its existence and geopolitical legitimacy as well as its function. In their work on an emerging culture of European diplomacy, Benson-Rea and Shore cite a European diplomat explaining to them during an interview how

> If you go into a country and say 'I'm the Ambassador of France', everybody thinks they know what an Ambassador is and what France is. If you say 'I'm the Head of the Delegation of the European Commission', they really haven't got the slightest idea what you're talking about. (Benson-Rea and Shore 2012, 480)

It is precisely this need for explanation of what the EU and its agencies really are, what they are there for, and what they do, that the regional integration narrative and the civilian/normative power debates are essential for legitimising both the EU's internal and external actorness. Drawing on Bourdieu, Kuus (2014, 164) argues that 'a claim of authority, however, legitimate, must be socially accepted before it is made. A claim of authority works best – which means that it goes unnoticed – when it is based on the claimant's position in the field of power'. The EU's position in the field of power is based on (the perceived) superiority of its model of political-economic organisation. I have tried to illustrate here the irony of the simultaneity of two key events/processes in this context. On the one hand, the Lisbon Treaty and the EEAS as institutional foundations for a more active and influential global role of the EU, both of which are a result of a phase of EUphoria in the 2000s and of a historical positioning of the EU as a 'better' geopolitical actor. On the other hand, the emergence of the Eurocrisis and its undermining of the narrative of functioning regional integration and the superiority of the European model *vis-à-vis* others. As such this chapter has provided the wider political and economic contexts within which the empirical data acquisition has taken place. Through a critical geopolitical approach, the second part of this book then studies practical geopolitical agents, exploring how, in light of these contexts, the EU is positioned and perceived as a global and development actor in a key node of African-European relations and the global development industry – the Kenyan capital of Nairobi.

Chapter 3
External Dimensions of Europeanisation and Regulated Spaces of Interaction

In the previous chapter I have addressed aspects of the discursive construction of European integration and the EU's identity and role in the world. In this chapter, I argue that these discourses, and more generally EU (international) agency, are inherently spatial. I situate European geopolitical discourse in wider debates on the spatiality of the political and the political nature of space, arguing that space is always socially produced and politically instrumentalised. Academic work in geography's subdisciplines of political geography and geopolitics has always centrally focused on the relations between space and the exercise of political power. Also critical geopolitics takes an interest in these relations, focusing on their social construction. Simultaneously, through its geographic interest in space, critical geopolitics was also part of a wider spatial turn in the social sciences and thus brought attention to spatial configurations to political science and international relations. Concerned with political spaces and the spatiality of the (geo)political, geographers (and others) have long explored the reciprocal influence between spatial configurations and political power (Harvey 1969, 1973; Lefebvre 1974; Raffestin 1977; see also Dell'Agnese 2008). Critical geopolitics was thus a decisive factor in bringing together the approaches of thinking (geo) politics both discursively and spatially.

It is in such a context that I introduce here the notion of spaces of interaction as a broad framework that governs the interaction of various political, economic and social actors. For the EU, the spaces of interaction that have developed through half a century of European integration are mostly regulated and institutionalised. Moreover, I also provide a brief geographical interpretation of the literature on Europeanisation as a key set of debates in European Studies. Geographers have only recently started to engage with the 'under-researched scales and spaces of Europeanisation' (Clark and Jones 2008, 301), highlighting how Europeanisation processes are spatially conditioned and how, reciprocally, they influence the making of European space. Exploring the relations between territory, scale, government and power, Clark and Jones argue that 'Europeanisation is a uniquely geographical suite of processes springing from territorial propinquity, comprising myriad socialisation and learning processes that have been configured over centuries by distinctive patterns of European government and power' (ibid., 313). This chapter draws on such pioneering work (see also Jones and Clark 2010; Moisio 2011; Rovnyi and Bachmann 2012; Moisio et al. 2013), employing a critical geopolitical lens and thus conceiving of Europeanisation as geopolitical practice and discourse. I argue that a key objective of such practice and discourse is the regulation of interaction whereby the EU 'seeks to turn "less European" spaces into fully European ones, both within the EU and beyond' (Moisio 2011, 35). Like the civilian and normative power debates, Europeanisation also has a certain missionary ambition with respect to the EU's external relations. Due to its experience in operating in Europeanised and regulated spaces of interaction internally, the regulation of interaction spaces is also a key objective of EU external relations. To introduce these debates I point to a differentiation between *Europeanisation abroad* and *external Europeanisation* and criticise a general lack of a critical reflexivity on the part of the EU when it comes to the external projection of internal interaction spaces.

Spaces of Interaction as a Geographic Framework for Political Agency[1]

> Space is an invention or, if you prefer, a construction; it is a concept that permits us to act. It is a question of nothing other than a tool modifiable according to the dictates of need. (Raffestin 2012, 123)

> space is an integral element in the constitution of everyday life and its associated social processes or, put differently, everyday life is constituted in and through temporal-spatial social relations. (Swyngedouw 1997, 143–4)

Political Spaces, Spatial Politics

Geographers have long engaged with the political nature of space and the spatiality of the political (see, for instance, Harvey 1996; Dell'Agnese 2008; Belina and Michel 2011b). Particularly influential for the development of a more thorough understanding of the relation between social and political processes and space has been the work of Henri Lefebvre. The 1991 English translation of his 1974 *La production de l'espace* triggered a more thorough engagement with the spatial dimensions of society, politics and the economy in Anglophone social sciences. Lefebvre was not a geographer but a sociologist and his prime concern was initially not with space. As a key figure of the French intellectual left and in the 1968 urban struggles in Paris, he was more concerned with processes of social exclusion and biopolitical governance and how these processes were spatially conditioned. Space, for Lefebvre, was therefore less interesting per se, but becomes of central relevance as *the* means and subject for the state to govern accumulation and reproduction. To understand those contested processes it is necessary to engage with space and its production (see also Lefebvre 1970; Lefebvre 1973; 1991 [1974], 281). Lefebvre understands space in terms of materiality, meaning and 'lived space' as a product of social practice: 'itself the outcome of past actions, social space is what permits fresh

---

1 This sections partially draws on Bachmann (2015b).

actions to occur, while suggesting others and prohibiting yet others' (Lefebvre 1991 [1974], 73). Space thus neither exists independently and beyond society nor as a pure idea void of connections to mundane materiality. As such, it is not space itself that is of interest, but its role in social practice and how it shapes (and is shaped by) social, political and economic relations (Belina and Michel 2011a, 17–18).

In *La production de l'espace* Lefebvre (1991 [1974]) offers, in his own words, 'an exposition of the *production of space*' (404) and claims that 'each [space] is *produced* – and serves a purpose; and each wears out or is consumed, sometimes unproductively, sometimes productively' (403). He terms such space, produced by the state, as 'abstract space' and argues that it is inherently political; 'instituted by a state, it is institutional' (285), whereby the state is conceived of as a 'spatial framework' (281). In a recent engagement with Lefebvre, Brenner and Elden (2009, 359) thus interpret abstract space as a 'political product of state spatial strategies – of administration, repression, domination and centralized power'. They continue that, for Lefebvre,

> the production of abstract space entails transformations not only in political practices and institutional arrangements, but also in political imaginaries: it involves new ways of envisioning, conceiving, and representing the spaces within which everyday life, capital accumulation, and state action are to unfold. (Brenner and Elden 2009, 359)

Strongly influenced by the work of Lefebvre, the spatial theorising of Claude Raffestin focused on territoriality, yet he was equally concerned with the influence of spatial configurations on the organisation of (political) life. Raffestin has early conceived of territory in relational and dynamic terms, arguing that 'the language of territoriality is that of relations' (Raffestin 1977, 125)'. Territoriality, to him, is 'not static but dynamic' (ibid., 132; author translations) and therefore a 'system of relations [that] is also a system of exchanges and, consequently, a system of flux of all sorts between exteriority (the physical environment) and alterity (the social environment). Territorial morphology is not explicable outside of the activities that created it' (Raffestin 2012, 129). Raffestin's entry point was thus a different one from Lefebvre's, for while the sociologist Lefebvre (1991 [1974], 281) considered the *state* as a 'spatial framework', the geographer Raffestin (2012, 123) identified *space* as 'a concept that permits us to act'. Nevertheless, both were concerned with the relation between space and political agency/power and viewed space as both socially produced and politically instrumentalised to enable the implementation of certain policies and disabling others (for related early debates in Anglophone geography see Harvey 1973; Smith 1984; Harvey 1989).

For political geographers, this relation between space and power is at the centre of their research interests. Emerging from this interest in the late 1980s, critical geopolitics started to engage more thoroughly with notions and assumptions of geopolitical space and the spatiality of international political-economic organisation. It is now a heterogeneous school of thought building on and exploring different foundations, sites and agents (Dodds, Kuus, and Sharp 2013; Moisio 2015). The critical geopolitical approach employed here emphasises the social construction of 'geopolitics'. It understands such social construction as historically and spatially conditioned, as framed by political and economic factors, as locally and internationally embedded, and as materially and representationally constituted (see Introduction). Accordingly geopolitical space is not demarcated by national boundaries, but 'implicated in the ongoing social reproduction of power and the political economy' (Ó Tuathail and Dalby 1998, 2). It is malleable and evolving over time through social, cultural, political and economic practices. Raffestinian thought is easily observable in these debates, be it the social or political construction of territoriality, and space more generally, or the dynamic and relational condition of (geopolitical) space. Throughout the 1990s, the spatiality of the political, the political nature of space, has been widely elaborated – along with a widespread acceptance of space as dynamically evolving and as a geopolitical factor and unit of analysis that is actively shaped by humans. Based on such an understanding of geopolitical space, I examine here the interplay of the factors of social construction of geopolitical space mentioned above, focussing on the creation of the interaction spaces for the making of EU diplomacy in Nairobi.

Relational Sociospatiality and Interactions

Relational approaches have been key to the evolution of spatial theorising – often taking Harvey's (1973) pioneering work on absolute, relative and relational space as an entry point (Massey 2004a, 2004b). As part of such relational thought, human geographers engage with relations between objects and actors in space and theorise different kinds of 'spaces', emphasising their relationality (Amin 2002; Peck and Tickell 2002; Elden 2005; Harvey 2006; Agnew 2011). Relational thinkers, according to Jones (2009, 492), 'replace topography and structure-agency dichotomies with a topological theory of space, place and politics as encountered, performed and fluid' whereby 'all non-relational thinking is conveniently displaced into notions of static space' (ibid., 494). They emphasise the fluidity, and evolving, malleable nature of relational space *vis-à-vis* a more static spatial understanding of fixed territoriality. Relational thinking thus rejects 'forms of spatial totality' and dissolves 'the boundaries between objects and space'. Space 'does not exist as an entity in and of itself, over and above material objects and their spatiotemporal relations and extensions. In short, objects *are* space, space *is* objects' (Jones 2009, 491).

Space has thereby largely ceased to be an exclusive category of analysis. Geographers have moved to more inclusive and comprehensive approaches that better reflect spatial dynamism, such as spacetime (Massey 1999), event-spaces (Marston, Jones, and Woodward 2005) or sociospatial relations (Jessop, Brenner, and Jones 2008). These approaches generally reject spatial staticism and instead accept space as relational, focusing on a process-oriented analysis of sociospatial relations – i.e. they examine space in its relations with society and accept a general reciprocal influence of society on space and of space on social (and political and economic) life. Interactions are at the heart of these relations between different actors in space and between space and those actors. They are at the centre of the practices and processes of 'perpetual becoming' and the mutual constitution of objects and space, space and objects that Jones refers to in the quote above. Also for Massey (2004a, 8) '"global space" is no more than the sum of relations, connections, embodiments and practice'. Massey was amongst the first to recognise the importance of interactions to these processes; both time and space, she suggests, should be conceived as 'the product of interaction, of interrelations' (Massey 1999, 274). It is because of this general acceptance of relational sociospatiality, as the key category of spatial analysis and the centrality of interactions to these approaches, that I suggest the term 'spaces of interaction' for capturing the underlying spatial configurations upon which sociospatial relations unfold and the practices and processes through which such relational sociospatiality is enacted (Bachmann 2015b).

Spaces of Interaction

> There is a space of speech whose prerequisites, as we have seen, are the lips, the ears, the ability to articulate, masses of air, sounds, and so on. This is a space, however, for which such material preconditions are not an adequate definition: a *space of actions and of inter-actions*, of calling and of calling back and forth, of expressiveness and power, and – already at this level – of latent violence and revolt; the space, then, of a discourse that does not coincide with any discourse on or in space. The space of speech envelops the space of bodies and develops by means of traces, of writings, of prescriptions and inscriptions. (Lefebvre 1991 [1974], 403, emphasis added)

The term 'spaces of interaction' first appeared in Henri Lefebvre's 1974 seminal work *La production de L'Espace* quoted above,[2] yet has long been neglected in geographers' engagement with Lefebvre (see, for instance, Brenner and Elden 2009; Lefebvre, Brenner, and Elden 2009; Belina and Michel 2011a; Kipfer, Saberi, and Wieditz 2013). I therefore want to unpack the quote above in more detail. Lefebvre refers to a 'space of speech' to which 'material preconditions are not an adequate definition'. Space in this sense is thus a construct or platform for the articulation of discourse, not (only) a physical condition. It is a construct that can be both physical and not. He further hints at its malleability and the process of shaping space through 'actions and interactions', through negotiations and 'calling and calling back and forth'. Lefebvre posits a space of 'expressiveness and power and […] of latent violence and revolt' that does not 'coincide with any discourse on or in space' because it is *the* space of a discourse. Discourse and space are not separate. Discourse is embodied in space and space in discourse. Space sets the

2 In the original as 'espace d'actions et d'inter-actions' (Lefebvre 1974, 463).

framework for discourse and discourse shapes space. Yet space is more than just a physical demarcation within which discourse evolves. It is a broad ideological, philosophical and political framework based on certain spatial understandings that regulate the interaction of objects and actors, of society and people and that, in turn, is influenced by how these entities actively shape this (spatial) structure for their interaction. In this context, Lefebvre has given an early indication of what Marston et al. (2005, 424) describe as 'the intimate and divergent relations between bodies, objects, orders and spaces'. He emphasises the human element and the power of discourse in society for shaping space. It is here that Lefebvre lays the foundation for accepting the *reciprocity of sociospatial relations* – the widespread assumption of the reciprocal relationship between society (shaping space) and space (influencing societal interaction) based on an understanding that 'spatiality is imbued with power *and* power is intertwined with spatiality' (Yeung 2005, 44).

This 'intertwinement' of spatiality and power is central to Lefebvrian conceptions of space as socially produced and politically instrumentalised. Both are inextricably linked with each other and with interactions. Social production emerges through interactions between social (and political) actors because of these actors' preferences for specific spatial configurations. Political instrumentalisation results from conscious social production of space in a process of making preferred spatial configurations work towards certain (political) ambitions/objectives. The arena (both physical and in terms of the regulatory framework) within which these processes occur, is what I refer to as spaces of interaction. Depending on the situation and the context, these spaces of interaction vary, take different shapes and operate in different ways. They are context-specific, evolving and conditioned through historical developments. They are the 'space between' (Bradbury and Lichtenstein 2000) and influence 'what happens in between' (Deleuze 1995) – i.e. the interactions through which sociospatial relations are enacted in various settings and multiple forms. Spaces of interaction thus constitute a (physical and regulatory) framework upon and through which political, social and economic life unfolds – 'a concept that permits us to act' (Raffestin 2012). Shaping, and being shaped by, the practices and processes of enacting (power) relations within society and between society and space, they are at the core of sociospatial relations. The arguments laid out in the following use the notion of spaces of interaction as an overarching framework for studying 'the spatial and temporal depth of [the] relations' (Degnen 2013) through which EU diplomacy in Kenya is enacted.

Regulating the Interaction Spaces[3]

> A way forward is to rephrase the question of what is the spatiality (or more narrowly, perhaps, the political geography) of EU governance towards critical scrutiny of how this is discursively constructed. Therefore, there is no single,

3 This section partially draws on Bachmann (2015a).

> stable, hegemonic understanding of the EU. Instead, a variety of visions exist in circulation and contest, embodying (though rarely in a straightforward or direct way) different social interests. (Sidaway 2006, 10)

After more than six decades, the process of European integration has produced a particular type of interaction space. As indicated by Sidaway in the quote above, this spatiality is (at least partially) discursively constructed – in debates such as civilian and normative power – and inextricably linked to 'EU governance'. Spatiality is dynamic and polymorphic and produces equally dynamic and heterogeneous 'understanding[s] of the EU'. In line with the arguments laid out in the previous section, such spatialities are inherently political, socially produced and political instrumentalised, and as such are tied to the exercise of political power. The integration discourses of the 1970s took a networked, connected, spatiality for granted without ever conceptualizing it. In 1972, François Duchêne suggested that Europe should collectively position itself as an alternative geopolitical model to the then two superpowers of the Cold War; a more civilian geopolitical model (Duchêne 1972, 1973). In Duchêne's vision, (Western) Europe should 'domesticate' relations between states – international relations should be dealt with in terms of domestic policy, i.e. regulated by a binding legal system and commitment to institutions. As such, Europe should set an example for a 'civilianized' international system based on its collective development from a war-torn continent to a political-economic system in which interaction is regulated through a common frameworks instead of the use of force (Duchêne 1973, 19). The functioning of that framework is to be based on the rule of law, multilateralism, supranationality and international institutions, a commitment to social and democratic standards, and the restriction of the use of force as a means for international politics (see also Maull 1993; Kirste and Maull 1996). Europe's influence on the international system would then be exercised through *functional* spheres of influence, as in trade, instead of primarily *territorial* ones (Duchêne 1972, 38). The focus therein lies on economic, diplomatic, and cultural exchanges rather than in military ones:

> a primarily civilian power on the scale of Western Europe […] could play a very important and potentially constructive role. More and more, security policies today, even for the super-powers, consist in shaping the international *milieu* often in areas which at first sight have little to do with security. (Duchêne 1972, 43)

As the geopolitical debates on European integration moved on, they retained a focus on shaping and institutionalising an 'international milieu' instead of territorial, realist fixation. More recent debates also take it for granted that the international system is organised through legal structures that create spaces for institutional actors, such as the EU, to exercise geopolitical agency (Gehring, Oberthür, and Mühleck 2013; Richard and Hamme 2013; Telò and Ponjaert 2013). Manners (2008, 45), for instance, argues that the EU 'changes the norms, standards and prescriptions of world politics away from the bounded expectations of state-centricity'. Contemporary debates retain the assumption that the EU's internal geopolitical space is historically and socially produced – a result of the experiences of the world wars, the Cold war and the European integration process – and that it should also be projected externally. Manners (2010, 83) talks of an 'ideological projection of the past onto the present', and for Rogers (2009) the civilian power strategy was a deliberate attempt to position a new, 'better' Europe *vis-à-vis* its chronopolitical opponent: its own war-torn past. A certain type of geopolitical space is thus being purposely developed internally and also positioned as a key geopolitical objective. The aligned debates have thereby almost always been both 'descriptive and prescriptive' (2002, 770). They describe how the EC/EU came to operate through 'functional spheres of influence' (Duchêne 1972), developed a 'civilianized international system' (Kirste and Maull 1996) and at the same time idealised these as the 'preferred world order model' (Hettne and Söderbaum 2005, 538).

To develop an understanding of the production, instrumentalisation and external projection of such geopolitical space, it is of central importance to examine the historical processes of European integration and its discourses. Following the descriptive/prescriptive nature of the integration debates, their spatial assumptions have been particularly relevant for this evolving process of constructing a European version/vision of geopolitical space – both internally as well as externally. 'Different visions and theorisations of the EU', according to Sidaway (2006, 10) are not only 'productive of what they purport to describe, but […] thereby [also] invite and invoke different actions and spaces of possibility for different subjects'. Through the European integration process, a civilianised system has largely materialised because Europe(ans) produced it in an attempt to overcome the disaster of two world wars. It embodies those historical experiences, was influenced by the geopolitical constellation of the Cold War and has been shaped by internal political and social preferences for the future of the EC/EU. The European polity thus exemplifies core aspects of civilian power debates, in particular the willingness of nation-states to cede sovereignty in favour of supranational integration. Moreover, policy conduct and spaces of interaction within the Union are highly multilateral and institutionalised as well as based on the rule of law and values which have been commonly agreed upon. Formal trade barriers within the Union have long been abolished and the use of military force is virtually irrelevant as a means for international politics within the Union. Within the EU, the spaces of interaction thus became domesticated, civilianised and *regulated.*

This historical development and Europe's relatively limited ability to project military power influenced the debates on, the visions and the reality of, collective EU external relations. The most recent treaty governing the EU, the Treaty of Lisbon (EC 2007b, Article 10–11), is rather explicit in this regard:

> The Union's action on the international scene shall be guided by the principles which have inspired its own creation, development and enlargement, and which it seeks to advance in the wider world: democracy, the rule of law, […] human

> rights, the principles of equality and solidarity, and respect for the principles of the United Nations Charter and international law.

The lead discourses – civilian and normative power – have thus had an enormous political impact, underlining their dual descriptive/prescriptive character (see Chapter 2). After the experience of developing, shaping and materializing a 'polity and space of connection, movement and interdependence' (Bachmann and Sidaway 2009b, 94), *regulated spaces of interaction* became not only intrinsic to relations between different actors within the EU but also to the conduct of its external relations. Successively regulating internal spaces of interaction has been fundamental to the European integration process for more than half a century. Consequently, by now this is a realm where the EU has extensive experience. It is a large bureaucracy and acts as such; it has developed a profound expertise in shaping and influencing the institutions, regulatory mechanisms and legal frameworks that govern interaction. It operates through networks of fixed and standardised arrangements that were agreed upon by its constituents as the preferred *modi operandi* for interaction with each other (i.e. the Treaty of Lisbon). For establishing such regulated spaces of interaction the EU thus also has a head-start compared to other players in the international system and consequently seeks to promote them. Holden (2009) argues that through its external relations the EU aims to 'develop structures ultimately favourable to Europe' (14), a mode of 'global governance conducive to its model of operation and its values' (17) in an attempt to externally project its internal spaces of interaction.

By contrast, the EU cannot operate in an environment of arbitrary and ad-hoc interaction, anarchy in the international system, and realist power politics, because it lacks the flexibility to spontaneously react, let alone to act, and the tools to project power in such an environment. Geopolitical space, configured in such often static-territorial terms erodes EU geopolitical agency (see the very limited collective European role in Ukraine and Syria, for instance). As the double descriptive/prescriptive nature of EU geopolitical discourse, regulated spaces of interaction also have a double function. They are the means through which the EU projects space *and* the end to be achieved through the projection of European space. Jensen and Richardson (2004, 5) recognise this double function and point out that 'European space itself is the subject of a fledging policy field which seeks to create a vision for its future, and to impose order on actions within it'. Regulating and institutionalising interaction spaces is thereby a key means *and* objective of European geopolitical agency, both for internal interaction as well as for the conduct of external relations. Regulated spaces of interaction are at the heart of EU spatiality. The conduct of international relations is organised in a predictable way, governed by a legal framework and routinised relations between different actors – ideally along the lines of the EU's internal system of political-economic organisation, which simultaneously serves as empirical proof of the functioning of such a system – at least before the crisis. In the following, I will argue that what is commonly described as Europeanisation is, in essence, a process of regulating the spaces of interaction both within the EU and beyond.

Europeanisation as Spatial Regulation

> Europeanisation is a legitimising process through which the EU strives to gain meaning, actorness and presence internationally. (Jones and Clark 2008, 546)

Since the foundation of the European Union in 1992, a broad research field in the social sciences has developed around the term Europeanisation. Very broadly and simplistically, this body of literature is concerned with analysing how European integration influences practices and processes of political, economic and social interaction in member states, on the European level, and more recently also with regard to EU external relations. Traditionally it focused on the 'domestic adaptation [of policy processes in member states] to European regional integration' (Vink and Graziano 2008, 7). Europeanisation, however, is far from being a clearly defined phenomenon or concept; it is defined and approached differently amongst various social science disciplines (Delanty and Rumford 2005; Clark and Jones 2008; Exadaktylos and Radaelli 2009; Conway and Patel 2010; Flockhart 2010; Jones and Clark 2010; Wong and Hill 2012; Moisio et al. 2013). For the remainder of this study I want to pursue a broad approach to Europeanisation which I have subscribed to elsewhere together with a colleague. Drawing on other geographers work, Rovnyi and I view Europeanisation as 'a dynamic, actor-centered, mobile, open-ended networked process, more subtle than the binary of national-supranational identity that remains embedded in many political science studies' (Clark and Jones 2009, 201). Rovnyi and I continue by framing Europeanisation 'even more broadly and fruitfully [...] as a historically contingent and socially contextualized practice that transcends a narrow view on high politics to include no less political areas of popular culture and mundane lived experiences' (Rovnyi and Bachmann 2012, 261).

In laying out our argument, we suggest a number of aspects where, in our view, mainstream Europeanisation research falls short and in which geographers can make valuable contributions to the debates (Rovnyi and Bachmann 2012). We acknowledge the contributions of political science and international relations research on Europeanisation, however, we also criticise a neopositivist attitude in such work focussed on hypothesis testing around different variables. Instead, we suggest employing geography's conceptual and methodological flexibility for the study of Europeanisation. Inspired by other geographers' pioneering work, we suggest a critical geopolitical take on Europeanisation as geopolitical discourse and practice (Jones 2006; Moisio 2011) as well as sociospatial examinations of the myriad and polymorphous interdependencies between space and power in different Europeanisation processes (Clark and Jones 2008; Jessop, Brenner, and Jones 2008; Jones and Jessop 2010). We are widely sympathetic with Nigel Thrift's call to study the 'little things' of geopolitics (Thrift 2000), Martin Müller's take on critical geopolitics not only as language but also as practice (Müller 2008) and Merje Kuus' place-based field studies on the European quarter in Brussels (Kuus 2011a, 2014). We therefore call for a more grounded and localised analysis of Europeanisation practices and processes in different time-spaces (Dahlman and Ó Tuathail 2005; Ó Tuathail 2010). The approach to Europeanisation employed here is thus based on the understanding of geopolitics outlined in the Introduction, i.e. as socially constructed, historically and spatially conditioned, politically and economically framed, locally and internationally embedded as well as material and representational. It is in such a context that this book offers an analysis of the diplomatic microcosm of enacting an emerging collective European diplomacy in Nairobi.

This microcosm, at the time of research following the implementation of the Lisbon Treaty and the establishing of the EEAS, is thereby considered as one such time-space of Europeanisation. The shaping of regulated spaces of interaction elaborated on in the previous section is seen in the wider context of the Europeanisation debate – or rather vice versa: Europeanisation is seen as a practice and process of shaping regulated spaces of interaction. This is part of the reciprocal relationship between the exercise of political power and underlying spatial configurations, shaping and shaped by politics and historical, social, economic and cultural dynamics –widely referred to as sociospatial relations. These sociospatial processes of Europeanisation occur as much within the EU as they do outside it. European integration, EU geopolitical discourse, Europeanisation and the aligned striving for regulated interaction spaces have never been limited to the territory of the EU (or the European Community before). Instead these processes always had an external component and were consciously promoted through EU external relations.

In the following I want to make a distinction between two spaces of interaction and outline a differentiation between what I will refer to as *Europeanisation abroad* and *external Europeanisation.* The former is restricted to the interaction of European actors and discusses how EU member states and institutions 'Europeanise' *their* external relations. The latter describes how interaction with *external* partners becomes 'Europeanised' and how 'European' spaces of interaction are projected onto third countries. As such this differentiation is also linked to the two steps of diplomacy outlined in Chapter 1. *Europeanisation abroad* refers to processes in the first step of diplomacy, i.e. foreign policy coordination amongst different European diplomatic agents, while processes of *external Europeanisation* instead takes place during the second step, i.e. the conduct of collective foreign policy towards third countries.

Europeanisation Abroad and External Europeanisation

Europeanisation abroad describes practices and processes of foreign policy Europeanisation and, more generally, aligning the external relations of EU institutions and member states. This applies to both the domestic formulation of foreign policy as well as its implementation abroad through delegations and embassies. Most established in this context is the literature on foreign policy Europeanisation (Wong and Hill 2012) with its traditional focus on examining the 'acceptance and incorporation of the European dimension in national foreign policy' (Michalski 2013, 885). My understanding of *Europeanisation abroad*, however, goes beyond such a narrow definition; it is less focused on structure and more concerned with actors and agency; it is also less interested in the national foreign policies of the member states and more interested in the emergence of a *collective* European foreign policy implementation through the newly formed External Action Service. Despite the differences of my understanding of *Europeanisation abroad* to earlier and internal accounts of Europeanisation, I share Moumoutzis' (2011, 625) rejection of the assumption that 'foreign policy change produced by Europeanisation constitutes a sui generis phenomenon that requires *ad hoc* explanation'.

A useful attempt at a more nuanced definition of foreign policy Europeanisation is Wong's (2008, 325) distinction between three dimensions: a) a top-down process of national adaptation and policy convergence (downloading or EU-isation); b) a bottom-up and sideways process involving the export of national preferences and models (uploading of national projection); and c) the socialisation of interests and identities, or 'identity reconstruction' (crossloading). The first two of these dimensions are perhaps the most relevant to the study presented in this book; with the first meaning as the most traditional one, appearing mostly in context of aligning member states' foreign policies with a previously agreed common foreign policy agenda. Moreover, it is also relevant for *external Europeanisation* with respect to the external projection of internal spaces of interaction. Wong's second dimension, however, is equally important as regards the formation of collective diplomatic agency and *modi operandi*. Given that the EEAS is a nascent institution adding a diplomatic component to a diplomatically inexperienced European Union, it has to draw substantially on the diplomatic expertise and experience of its member states. In so doing, *Europeanisation abroad* also includes the formation of a collective European diplomatic culture (Benson-Rea and Shore 2012) and an *Esprit de Corps* in the EEAS (Juncos and Pomorska 2013b) through the process of uploading (and crossloading) diplomatic experiences of the member states (see above and Wong 2011, 158).

Most of the work on *external Europeanisation* has focused on the EU's exercise of structural power towards the Eastern European countries (see, for instance, Schimmelfennig and Sedelmeier 2005; Holden 2009). According to Boedeltje and Houtum (2011, 130), Europeanisation is thereby 'explained by the EU as a normative process of sharing European values made concrete through policies of conditionality and socialisation of neighbouring states'. Here conditionality is understood as the 'use of positive incentives to alter a state's behaviour' (Kelley 2004, 37) and has been employed by the EU with respect to its neighbourhood as well as towards African countries. The 'positive incentive' for pre-accession countries is, of course, the possibility of eventual membership; towards African countries, it refers predominantly to development aid or specific trade agreements. The second strategy, socialisation, is employed in the 'absence of material incentives [whereby] actors internalise EU rules and norms because they perceive them as appropriate and legitimate. Socialisation takes place through arguing and persuasion and might lead in the long term to changes in interests and identities' (Juncos 2011, 373). Recent examples of studies on *external Europeanisation* include, for instance, Juncos' (2011) work on policy reform in Bosnia, Petrov and Kalinichenko's (2011) engagement with the Europeanisation processes of third countries judiciary systems, Langbein and Wolczuk's (2012) research on the mechanisms through which the EU influences Ukrainian actors towards selecting, adopting and applying EU rules or Hulse's (2014) study of how other regional organisations adopt European spaces of interaction to develop trade actorness.

Whilst the majority of this is concerned with the EU's immediate vicinity, a comprehensive special issue in the journal *West European Politics* (2012, Vol. 35: 1) also examines broader practices and processes of *external Europeanisation* far beyond the EU. The editors of the issue explain how 'students of Europeanisation have identified both necessary conditions ("misfit" giving rise to adaptational pressure) and causal mechanisms (legal imposition, positive and negative incentives, and socialisation by persuasion and learning) through which "Europe hits home"' (Börzel and Risse 2012a, 1). Simultaneously, they criticise such approaches as hierarchical, as often falsely identifying the EU as the cause for change in third countries and as decreasingly relevant with increasing distance to the EU (ibid., 2). To overcome these shortcomings they suggest combining Europeanisation research with diffusion studies. Europeanisation, they argue, should not be overstretched as a concept and instead diffusion literature should be included to 'also capture the more indirect ways in which the EU influences domestic institutional change in third countries and regions' (Börzel and Risse 2012b, 193). The special issue illustrates a wide range of examples of processes of Europeanisation and diffusion of European policies and institutions to different countries and regions (Alter 2012; Jetschke and Murray 2012; Lenz 2012; Van Hüllen 2012).

These various case studies are of interest for their empirical orientation examining how EU policies spread outside the EU. Yet their conceptual grounding in diffusion studies sits uneasily with the critical geopolitical approach of this book. This approach focuses on the social construction of (geo)politics and accounts for its local embeddedness as well as the specific spatial and historical contexts. Instead of a mere diffusion, or rather a direct transfer, of a policy regime, such policy regimes as Europeanisation mutate when travelling and being implemented elsewhere (McCann 2010; Peck and Theodore 2010). Peck criticises how the diffusion paradigm, 'operating in relatively denuded policy-making spaces is seriously limiting, and leaves largely unexamined the *social* processes of policy formation and the context-dependent setting in which "borrowing and learning" take place' (Peck 2011, 791). Instead he suggests a 'social-constructivist understanding of policy mobilities-and-mutations, sensitive to the constitutive roles of spatiotemporal context' (ibid., 773). The attention to such 'spatiotemporal context' is indeed essential to the critical geopolitical approach employed here and to practices and processes of how collective European diplomatic agency is constructed in Kenya. Yet despite these conceptual differences to the diffusion paradigm advocated by Börzel and Risse, I share their empirical interest in the external projection of EU models. What is important to acknowledge here is how the 'adoption of EU institutional models' (Börzel and Risse 2012b, 194) in third countries or regions is not coincidental but actively pursued through practices and processes of external Europeanisation (including mutation).

Accounting for this 'active pursuit' of Europeanisation outside the EU, I maintain the differentiation between *Europeanisation abroad* and *external Europeanisation* in this book. Whilst the former remains limited to interaction between European actors, the latter refers to the active projection of 'European' practices and processes to the interaction with "others". In light of Europe's imperial past, it therefore has to be scrutinized much more critically and reflexively, in particular as regards the EU's relations with African countries. In a critical account of the EU's external relations, Boedeltje and Houtum (2011, 140) even posit that '"Europeanisation" in this sense suggests a new colonisation mechanism presented by a new empire called the European Union, and in this sense is abducting Europe from its intended open meaning as a continent not strictly defined'.

Europeanisation, External Relations and the Question of Space and Power

If Europeanisation is indeed a 'new colonisation mechanism' remains questionable. Nevertheless, it is approached here in the wider context of the EU's attempt to regulate external spaces of interaction – both between European agents and with external cooperation partners. The 'process of expanding Europeanisation beyond the EU borders' Boedeltje and Houtum (2011, 130) argue, 'is inspired by an ambiguous and conflicting geopolitics that the EU applies as a strategic instrument' (2011, 130). This is part of a wider ambition to project the internal spaces of interaction externally in an attempt to create an international system along the lines of the EU's internally preferred *modi operandi* – a system in which the EU possesses the greatest experience and within which it can most influentially exercise (political) agency.

This reciprocal relation between spatial configurations and political agency, or as Virginie Mamadouh (2010, 320) puts it, the 'sexy combination of space and power' is of traditional and fundamental interest to (critical) geopolitical research. How does space or certain spatial configurations influence the projection of political power? In the case presented here both are mutually reinforcing. Power and the means of power-projection shape and instrumentalise space, and at the same time space instrumentalised according to a specific actor's preferences increases the power of the actor within this space. As for European integration, it became easier to shape internal spaces of interaction the further this process had advanced and the more these spaces of interaction became regulated. Focusing on spatial planning, Moisio and Luukkonen (2014, 14) speak of a routinisation of EU policy making, referring to

> the routinized ways of managing decision-making processes – that is, the procedures related to the evaluation of the consistency between the EU policy programmes and the EU-financed regional development projects. Perhaps, more importantly, however, routinization refers to the adoption of a certain way of doing things.

The EU also prefers such 'routinized ways' of decision-making and regulated spaces of interaction externally because of its history and the historical practices and processes of producing such spaces of interaction internally. The experience of war, in particular World War II, and the subsequent political and economic recovery, led to the development of civilianised and regulated spaces of interaction in Western Europe. Through this process, the EU (and its predecessor, the EC) has developed a profound expertise in networked, cooperative, multilateral governance. It produced respective spaces of interaction that it instrumentalised, both internally as well as externally, to advance its agenda, i.e. to further regulate the spaces of interaction. Such regulated and institutionalised spaces of interaction thereby function as a means and an end of EU geopolitical agency: as the framework through which the EU acts and which it tries to achieve through various policies. This is part of a wider ambition to establish predictability and calculation, to bureaucratise political agency (Elden 2007; Moisio 2015). It is in and through such regulated and routinised interactions that the EU can most effectively exercise power. Ad-hoc politics and anarchy in the international system render the EU marginal as a (geo)political actor.

The key geopolitical debates, civilian and normative power, have played a key role in advancing the regulation of interaction spaces, debates that have always been both descriptive and prescriptive, describing how the EU operates and at the same time how an idealised geopolitical system should look (see Chapter 2). The internal modes of interaction are the ones that form the basis for the 'visions and scripts' of the 'global political space' to be projected externally (Ó Tuathail 1996, 185). More concretely, and with a focus on the political geographies of Europeanisation, Moisio et al. (2013, 17–18) argue that

> it would also be crucial for further analyses to turn their attention to the 'globalizing' and 'transnationalizing' practices which have always been an integral part of European integration, thereby disclosing the differential operation of these practices (and their underlying logics) in various geographical contexts […] Europe and Europeanization are shaped by interactive processes not only within and among member states, but also in relations with third countries. Without engaging with the meaning of Europe for those outside it,

> it is impossible to grasp the full extent of Europeanization, which reaches far beyond the external borders of the EU.

Part of such an engagement with the 'meaning of Europe for those outside it' has to be an engagement with resistance and objection to such processes of external European (power) projection. What is often brushed over in accounts calling for an external projection of the European model is that European integration emerged in very particular historical and geopolitical contexts; it is therefore decisively different from political interaction on a global level and can therefore not easily be replicated elsewhere. An unreflexive approach idealizing the European model and deducting from such idealisation the right to external projection is therefore highly problematic – in particular in light of Europe's imperial history. In contrast, this book attempts to lay out a more reflexive and critical take on the external projection of internal interaction spaces through a critical geopolitical study of the everyday practices of building European diplomacy in Nairobi as a key hub of African-European relations and the global development industry. Before presenting the key findings in the second part of this book, Chapter 4 will provide an outline of the wider context within which this work's fieldwork has been conducted and of the research that lays the ground for this study.

Chapter 4
An Institutional Ethnography of the Sites of Diplomatic Conduct in Nairobi

In Chapters 2 and 3, I illustrated how the external projection of regulated spaces of interaction – tied to regional integration and civilian/normative power narratives – has long been an essential part of the EU's external relations. Before the Lisbon Treaty this external projection of internal interaction spaces mostly occurred in trade and development policy. Diplomacy, on the other hand, was a national policy field, conducted bilaterally by the member-states or intergovernmentally. This has changed through the establishment of the EEAS – and diplomacy became of interest to studies of EU external relations. I use here a critical geopolitical approach to the study of EU diplomacy. Conceiving of diplomacy as social practice instead of official scripts requires a research approach different to the study of texts. It requires more direct, immediate research methods, such as interviews, ethnography and participant observation. It also requires attention to detail and a thorough understanding of the places, contextualities and social dynamics within which diplomacy is enacted. 'Situating diplomatic practices explicitly in the places where they physically take place', Kuus (2014, 38) argues, 'allows insights into the interpersonal aspects of geopolitical knowledge production that remain overlooked in traditional accounts'.

This chapter explains my approach to how I developed such an understanding during the course of this project. It introduces the Kenyan case study and reflects on the fieldwork process of studying how the EEAS is positioned and perceived as the EU's diplomatic agency in Nairobi.

The Local Context: Post-independence Kenya and its Relations with the EU

Nairobi has been chosen as the site for this study due to its pivotal role for African-European relations. This site allows for focussing on the emergence of an EU *political* role in an environment overshadowed by the international development industry (see Introduction). The observations and analysis made here are of wider relevance for understanding practices and processes of making EU diplomacy, yet the national context under which the study took place is that of Kenyan-EU relations. I will therefore briefly introduce the political-historical background of Kenya and its relations with the EU.

Kenya's immediate post-independence phase has immediate ramifications for the political situation of the 2010s, not least because the fourth Kenyan president is the son of the country's independence hero and first president Jomo Kenyatta. Upon independence in 1963, Kenya developed into a centralised, de facto one-party system of Kenyatta's party, the Kenya African National Union (KANU), with extensive powers vested in the hands of the president. Vast amounts of the land and key state positions were distributed amongst Kenyatta's family and other members of the ruling elite, mostly of the Kikuyu community. After Kenyatta's death in 1978, the political climate under his successor Daniel Arap Moi became increasingly restrictive. Moi institutionalised the one-party system and built an excessive police and secret service apparatus, pursuing political opponents and perceived dissidents. With the end of the Cold War, 'Western' governments, no longer in need of an anti-communist stronghold in the region, increasingly criticised the repressive (and heavily aid-dependent) Moi regime. With growing internal pressure and under the international threat to withhold development aid, Moi agreed to a range of reforms, including a multi-party system and the limitation of the presidency to two terms. Citing his capacity to keep the country united in the face of ethnic divisions, Moi won the elections in 1992 and 1997, before stepping down at the end of his 'second' term in office (see Oyugi, Wanyande, and Odhiambo-Mbai 2003).

In December 2002, Mwai Kibaki came to power through an electoral victory against Uhuru Kenyatta, Jomo Kenyatta's son and the candidate backed by the ruling KANU party and the outgoing president Moi. Kibaki, from the community of the Kikuyu (like the Kenyattas), headed the multi-ethnic National Rainbow Coalition (NARC), supported by the country's second largest community, the Luo, and their leader Raila Odinga. The NARC campaign ran on the promise of overcoming ethnic divisions as well as the kleptocratic structures of previous regimes (Bachmann 2010, 3). The Kibaki/Odinga alliance sought to unite the country's largest ethnic groups against the 'old' political elites and was built on a Memorandum of Understanding (MoU) that foresaw constitutional reform, including political devolution and a restriction of the powers of the president through the creation of the post of an executive prime minister. In 2005, Kibaki abandoned his commitment to the MoU, strengthened his grip on power through maintaining Kenya's centralised structure and placing confidants in key political and economic positions. Odinga in turn abandoned the alliance and started to build a political platform, the Orange Democratic Movement (ODM), in order to challenge Kibaki during the next elections to be held in 2007. The failure of this MoU is now widely regarded as a key factor triggering the violence in the country in 2007/2008. As Maina and Muliro explain shortly after the elections in the Kenyan *Daily Nation*:

> The country experienced a peaceful transition of power in 2002 on an overwhelming mandate for the multi-ethnic National Rainbow Coalition, Narc. This support was driven by two main promises – that of revival of the economy and conclusion of constitutional reform. The collapse of the Coalition's

> Memorandum of Understanding (among the coalition partners) on the sharing of political power, and therefore economic goods, weakened the multiethnic and inclusive character of the Government. Now, it has had far-reaching implications on the winners and losers in the economic recovery in the past five years as well as the constitutional debate and conference.
>
> The issues that have surfaced during the raging post-election crisis bring to the fore latent tensions present in Kenyan society. Such tensions revolve around equal chances to earn a living and the ability to have in place inclusive governance systems that capture the aspiration of many in the nation. (Maina and Muliro 2008, 19)

The violence that erupted in the country in the wake of the elections thus brought to the fore unresolved problems Kenya had inherited from colonial rule and that had remained untouched by all of the post-independence governments. With Kibaki, as another Kikuyu president, disrespecting the commitment to the MoU, the dissatisfaction with the ruling elites surfaced in particular amongst the Luo communities. In this regard, a local BBC correspondent argued at the time that 'the incumbent government in Kenya is perceived by the opposition as being Kikuyu-dominated, so the current political dispute is fuel for the smouldering embers of a land dispute which has existed for decades' (Doyle 2008). Similarly, a range of commentators in the *Daily Nation* lamented a long-entrenched political culture of elite enrichment, positing that

> [o]ur people, from Mount Elgon in the west, to the Kenya Coast, have always asked for change, from the structures and policies that we inherited from colonialism. But our leaders, starting with Jomo Kenyatta down to Mwai Kibaki, have always refused to listen […] In Africa, power is used to develop the regions of the leaders […] Our leaders campaign to go to Parliament not to cater for the interests or concerns of their electors, but to find ways of enriching themselves. (Ochieng 2008)

It was in such a climate of hostility amongst the different communities that the violence erupted in the country following the 2007 presidential elections between the followers of Kibaki on the one side and those of Odinga, including at the time the Kalenjin community and their leader William Ruto, on the other. The country witnessed almost two months of fighting with the result of over 1100 people killed and 660,000 internally displaced (Amnesty International 2014). After the crisis, initial investigations on the backgrounds of the violence were conducted under the leadership of the Kenyan judge Philip Waki. One of the recommendations by the commission headed by Waki was the establishment of a local special tribunal in Kenya to try the main perpetrators. A report by the International Crisis Group recounts how

> mindful of the history of political impunity, it [the Waki Commission] recommended that if the government failed to establish the tribunal, the Panel of Eminent African Personalities that under Kofi Annan's chairmanship mediated the political crisis should hand over a sealed envelope containing the names of those who allegedly bore the greatest responsibility for the violence to the ICC [International Criminal Court in The Hague] for investigation and prosecution. (ICG 2012)

The commission's recommendations were supported by both Kibaki and Odinga, President and Prime Minister, in a power-sharing deal negotiated in late February 2008, and adopted by the parliament in January 2009. Yet the establishment of a local tribunal failed twice due to the parliament's refusal to pass a respective bill. Consequently, the envelope was handed to the ICC for investigation in July 2009. In December 2010, the ICC announced a list of six individuals identified as those primarily responsible for organizing and inciting the violence on both side. The list included Uhuru Kenyatta as the key figure behind the scene on the Kibaki side, and William Ruto as the key person responsible for the 2007 violence on the Odinga side (see ICG 2012). Following their opposition to being tried at the ICC, Kenyatta and Ruto joined into a coalition for the 2013 presidential elections, which they eventually won based on a campaign positioning the ICC as a neocolonial institution interfering in Kenya's domestic affairs. These developments had a profound impact on Kenya's relations with the EU and the building of a collective European diplomacy in Kenya during the first four years of the EEAS (see Chapters 5 and 6).

Kenyan-EU Relations

Relations between the integrating European polity and Kenya have a long history. Already in the immediate post-independence phase, Jomo Kenyatta had sought close relations with the 'West', pursuing a broadly liberal economic policy and an integration in international markets (Chazan et al. 1999, 263). The delegation of the European Commission opened in Kenya in 1976 and, in the same year, Kenya was the first country to sign the newly established Lomé Convention (EEAS 2014b). Since then, the overarching cooperation frameworks were part of the European Development Fund (EDF): first the Lomé and, since 2000, the Cotonou agreements which cover several areas of cooperation, mainly development and trade but also security, humanitarian aid and others. Economically, the EU is Kenya's largest trading partner; in 2011 it accounted for 17.2 per cent of Kenya's overall trade (with China, the second one, accounting for 11.7 per cent) (EEAS 2014c). In terms of development cooperation, the key sectors focus on achieving 'regional economic integration by means of transport infrastructure and agriculture and rural development' and through 'general and sector budget support [as] the main aid instruments' (EC 2007a, vii). As such, the Commission is the only international 'partner' to Kenya actively pursuing the heavily contested tool of budget support in its development policy (see Chapter 6).

Political relations have fluctuated over time, being subjectively dependent on the political climate at a given time and on perceptions about the compliance with conditionalities related to good governance, on the one side, and on resentments against European paternalism, on the other. The period during which this study was conducted was highly problematic and contested, overshadowed by the events that followed the 2007 elections and the pending cases of Kenyatta and Ruto at the International Criminal Court (ICC) for their alleged involvement in the post-election violence (see Chapters 5 and 6 in this book; also Kiamba and Bachmann 2015). Even though nominally the EU had no collective *political* role in Kenya prior to the Lisbon Treaty and the establishment of the EEAS, relations have always been power-laden due to the highly political nature of 'Development' (Ferguson 1990) and trade policy as defined in the Lomé and Cotonou agreements. In 2010, the political nature of these long-establishment relations was institutionalised and formalised in the form of the EEAS, adding an official political mandate to the EU's collective relations with Kenya. Prior to the EEAS, foreign policy towards Kenya was in the hands of EU member states that had been holding long and intensive relations with the Kenyan government since independence. Naturally, EU member states maintained their relations, also in light of an emerging collective diplomacy. On the one hand, the timing of the study thus allowed for observing the rivalry between national and collective diplomacy during the phase in which the relationship between the two was defined. On the other hand, it also allowed for observing practices and processes of 'uploaded Europeanisation' (see Chapter 3), i.e. the projection of national expertise to the European level. This was particularly important, since the EEAS as a nascent service had little political lobby and heavily depended on the member states for expertise, contacts, and cooperation – both for its formation as a diplomatic service but even more so as regards the interaction with the Kenyan government (see Chapter 5 and 6).

For a researcher, studying such an institution in the making is not an easy endeavour as the institution itself is still in the process of defining its identity and role – and often also of legitimising or even justifying its very own existence. By the time this research project started in 2010, the EEAS did not yet exist. It was part of the Lisbon Treaty that had just come into effect in December 2009, and preparations were underway for its official launch in Brussels one year later. My first trip to Kenya in the context of this project (November/December 2010) fell into the period of the launch of the EEAS. During the time of that research stay in Nairobi, even officials at the Delegation were unaware of the changes the EEAS was supposed to have already introduced and would be bringing to their work. An EU official in Nairobi openly admitted how 'the EEAS in Nairobi is not really happening yet. Nobody knows what it will be like. There are no instructions from Brussels for how it should be materializing in practice' (Author interview, November 2010). Given the uncertainty about the EEAS – even within its own staff – I encountered significant difficulties in the research process of trying to find out about how the EEAS is used by different European actors to collectively position themselves on the political level. Even more difficult were inquiries into how the EEAS is perceived amongst the EU's cooperation partners, such as the Kenyan government and its other local and international partners. My research thus commenced in a phase of uncertainty on all parts about that new institution. It was a phase of orientation for both its internal 'makers' and its external cooperation partners and, of course also, for researchers interested in this new institution.

Studying the Places and Social Dynamics of an Institution in the Making[1]

For approaching this emerging institution, I conducted several fieldwork trips to Nairobi and Brussels between November 2010 and March 2014, spending a total of seven months in Nairobi. In particular in 2010, there was a wide-spread lack of knowledge of and experience with the EEAS amongst both African and European diplomats and officials. At subsequent fieldwork trips between 2012 and 2014, this had changed significantly. Nevertheless, in light of the cluelessness shown by virtually everybody supposed to be 'making' or cooperating with the EEAS in 2010, an exclusive focus on interviews as research methodology appeared too shortsighted at the beginning of my project. I had to realise quickly that what people could tell me about the EEAS was often limited as they knew little about it themselves. I thus resorted to a multi-method approach, including ethnographic methods for 'deeply immersing' myself in the field (Gans 1982) as well as the study of both local and wider geopolitical contextualities.

I have previously laid out a critical geopolitical take on the study of international organisations (IOs) that also informs the work on the EEAS presented here (Bachmann 2013c). This approach moves away from a rationalist-institutionalist perspective that sees IOs as being granted agency by nation-states and tools for 'enabling nation-states to achieve their ends' (Abbott and Snidal 1998, 3), not actors sui generis. In contrast, Barnett and Finnemore call for a more sociologically informed, constructivist approach towards the study of IOs, arguing that the 'rational-legal authority that IOs embody gives them power independent of states that created them' (Barnett and Finnemore 1999, 699). The EEAS has been given such legal authority by the Treaty of Lisbon as the central institution for the conduct of EU external relations and has been created as an actor sui generis. To study the 'rational-legal' authority of IOs, Barnett and Finnemore (1999, 726) suggest paying more attention to their day-to-day operation, and as a result their research has been instrumental in triggering more sociologically oriented approaches in the discipline of International Relations with respect to the study of international organisations; research that is concerned with what happens inside the organisation and the respective social dynamics at play. Barnett and Finnemore (2004, 9) further state that 'by opening the black box of international organisations and examining how they are constituted and use their authority, we

1 This section partially draws on (Bachmann 2013c).

can begin to understand their power, their capacity for pathological behaviour, and the way they evolve'.

Martin Müller (2012), too, takes an approach to the study of organisations that calls for an opening of the black box of an organisation and examining what happens inside, and not only the output it generates. Müller's argument emerges in a wider context of his work that seeks to expand the Foucauldian approach to discourse analysis beyond language by suggesting to look at how meaning is constructed by signifying practices (see also Laclau and Mouffe 1985; Müller 2008). Attention to detail, to the mundane practices and interaction behaviour within an organisation is thus essential to capture its internal dynamics and for understanding how an organisation's external mission and agency is developed. John Law (2004) suggests a baroque view, one that looks down at the small, incoherent and complex that together constitute a whole instead of looking up in search of large-scale coherence. According to Kuus (2010b, 383), this baroque view 'directs us to look within, into specificities, uncertainties, and discrepancies that are brushed over in big picture accounts'. In the context of her work on EU policy-making procedures, Kuus therefore points to the need 'to investigate not the large systems but the small, the irregular, and the baroque; not to enumerate power struggles among the member states, but to study the mundane bureaucratic practices in Brussels through which these struggles are codified' (ibid.).

Whilst these power struggles are clearly also a relevant research endeavour, I want to reiterate Thrift and Kuus' call to pay more attention to the 'little things' (Thrift 2000) and the dynamics that unfold within an organisation and that form a – however loose – coherence for its external action. These dynamics depend more on people and their social relations with each other than they do on official doctrine or policy documents. Kuus further (2010b, 383) suggests to 'analyse the operation of a social alchemy whereby a social hierarchy dissimulates itself no less to those it dignifies than to those it disenfranchises'. Some individuals operate more smoothly and more efficiently than others, some have more experience in how the processes work, other have more expertise on a certain topic. This multitude of people with different backgrounds and interests as well as varying strengths and weaknesses is to be found in each international organisation. Despite the diverse composition of many IOs, they constitute an entity – *an* organisation – and not just an arbitrary agglomeration of individual components. An organisation is therefore more than the coexistence of its constituent parts, it develops agency through a 'pattern of recurring linkages among its parts' (Lincoln 1982, 26). Drawing on Callon and Latour (1981), Müller explains how 'organizations derive their agency by enrolling a multitude of humans and things and making them work towards a mission' (Müller 2012, 382).

Networked Places of European Diplomacy

Reflecting on approaches in sociospatial theory (Jessop, Brenner, and Jones 2008; see also Chapter 3), the EEAS is a geopolitical agent that operates on and through multiple spatial dimensions, including scale, networks and place. It works across a range of different scales. Internal processes of multilevel governance produce the EEAS as an institutional body in the EU system that is both supranational and intergovernmental. Yet, it is closely tied with foreign policy and the diplomatic conduct of the member states and towards the Kenyan government. Moreover, it also operates with civil society organisations as well as other international organisations and the UN system. Even though this scalar politics is inherent to an emerging system of EU diplomacy, the focus here is on the networked places where such diplomacy is enacted.

Place has long been a central category of geographic research (Massey 1994; Amin 2004; Cresswell 2004; Massey 2004a). The interdependencies between place and networks are also well documented, whereby places are described as nodal points in which wider political, social, cultural and economic networks are grounded and where they physically materialise (Amin 2002, 2004; Nicholls 2009; Pierce, Martin, and Murphy 2011). Places are, therefore, influenced by their transgressing networks and, of course, simultaneously influence the structures, processes and flows of these networks. Pierce et al. (2011, 67) argue

> that *all* places are relational, and are always produced through networked politics. Our intervention aims to help geographers by providing an analytical approach that helps us do more explicitly what we are already doing implicitly: to consider the interconnections and co-constituencies among place, networks and politics by identifying specific conflicts and the places they produce, the dimensions of place-framing evident, and the multiply-positioned actors and places/bundles inherent in and underlying the conflicts.

Such multidimensional analysis is relevant to the study of international organisations in general and the EEAS in particular given their wide-ranging international networks as well as their place-based social dynamics (Jessop, Brenner, and Jones 2008). An organisation's agency is often channelled and materialises through the networks of both the individuals working in the organisation and the more structural networks of the organisation itself. The EEAS, for instance, ventures into the established networks of EU member states with each other, with the Kenyan government and with other agencies present in Nairobi. It is fed by the expertise of former national and EU diplomats and officials. Those professionals retain the networks of their previous institutions to inform their work at the EEAS and build its expertise – but also to connect the internal developments of the EEAS with their former institutions. Networks never only operate unidirectional. However, these networks become enacted in physical locations, in nodes where practical and physical interaction materialises and where they are connected with others and other networks. The processes and flows through which networked actors (such as the EEAS) develop agency are therefore not purely determined by the structure of their institutional arrangement; they are equally shaped by the people, the social characters and their habits of interacting with colleagues. Viewpoints are articulated, compromises negotiated and decisions made by people and their interaction with each other. These negotiations, encounters and interactions are the nodal points of networks and are physically located in particular places.

In the case of EU diplomacy in Nairobi, a key place is the Delegation of the European Union to the Republic of Kenya; or, even more so, the meeting room of that building (for more detail, see Bachmann 2015b). Of course the EEAS headquarters in Brussels is also an example of a place where European External Action is enacted. Studying such sites of diplomatic conduct through a critical geopolitical lens, my research draws on Merje Kuus's work on policy-making processes and geopolitical knowledge generation in Brussels (Kuus 2011a, 2014). Kuus has insightfully highlighted how the social dynamics of particularly places are decisive for the geopolitical agency of the institutions that inhabit these places. Focusing on Brussels she argues that 'to highlight the small and the circumstantial is not to ignore the big picture of inter-state power struggles […] it is rather to substantiate the broad-brush explanations with a finer-grade analysis of daily work in the European Quarter' (Kuus 2014, 6). It is therefore of vital importance to understand how people interact with each other in their daily lives in order to understand how agency in an organisation is produced. The critical geopolitical approach employed here therefore seeks to trace 'the qualities of experience and meaning-making in everyday, in-between spaces, and the ways in which people make sense of the big world through the small' (Bendiner-Viani 2013, 708). This implies accounting for networks in order to develop an understanding of the contextualities within which an organisation operates, positions itself and is perceived by its partners. But it also implies accounting for places where networks connect in order to develop an understanding of the internal social dynamics that are decisive for decision-making about the organisation's external agency. It is through such a critical geopolitical approach, sensitive to multiple spatial dimensions as well as to both local and wider geopolitical contextualities, that my work studies the build-up of the EEAS in Nairobi (see also Bachmann 2013d). It does so by accounting for its internal dynamics as well as for its external position between its constituting elements (EU bodies and member states) and cooperation partners (Kenyan governments, NGOs, etc.).

Pragmatism in the Field

I follow hereby an approach laid out by political scientists Friedrichs and Kratochwil's (2009) that calls for employing more pragmatic research approaches modelled on 'real-world' experiences of knowledge generation. They argue for the need to conceive of 'knowledge generation as a social and discursive activity' (701) and to acknowledge that 'at the bottom of our hearts, we all know that the way we produce knowledge in our everyday social practice has advantages over standard scientific methods' (702). For the discipline of International Relations, they criticise that 'some scholars pursue the accumulation of law-like statements as if positivism had never been rocked. Independent, intervening, and dependent variables are tossed around as if the social world resembled a bowling alley' (Friedrichs and Kratochwil 2009, 702). Instead they argue that, in particular for social science, a pragmatic approach to knowledge generation is useful. The purpose of social science knowledge, according to Friedrich and Kratochwil, is to 'enable orientation in the social world. Its utility consists of helping us to understand complex social phenomena and/or to explain social regularities' (706). Consequently, the generation of respective knowledge is less abstract and should be oriented more towards the social world and the interaction of individuals and institutions within it. Understanding ourselves and arranging our position and life in this social world is thus a prerequisite for enabling us to generate knowledge about it. Social science, Friedrichs and Kratochwil argue, 'is, or should be, above all a more conscience and systematic version of the way by which humans have learned to solve problems and generate knowledge in their everyday lives' (710). These ways and skills of 'real world' problem solving are thus to be utilised for research and also acknowledged as such by the researcher. Employing pragmatism and common sense, has long been a practice for most researchers in the social sciences, yet subscribing to it as a philosophical and methodological position hardly ever takes place. Such subscribing, however, should not be understood as a call to abandon academic rigour in the conduct of fieldwork. Friedrich and Kratochwil (2009, 714) argue that

> [p]ragmatic researchers will not uncritically impose their own categories, nor will they become enmeshed in interpretive or ethnographic research to such an extent as to surrender all critical judgment to the practitioners in the "field". […] What we need, therefore, is a compromise between some rockbottom standards of scientific methodology and the way we produce knowledge in everyday social practice.

There is no standard of such 'compromise'. Each research environment is different and each researcher has to negotiate her/his compromise and relation with the field, the informants and the wider research environment her/himself. As part of these processes of negotiating one's position in the field and of becoming immersed in that field, ethnographic research often transcends the strict separation between research and non-research. It is a matter of gaining access to groups of people, to live and/or work with them, spend time with them, but also to study them. Evans (1988) suggests identifying certain significant individuals, gatekeepers, attaching oneself to them and building other field-relationships based on the relationship with the gatekeeper. Ideally the researcher then positions her/himself both 'inside and outside the group', thus being able to provide 'an insider's account with an outsider's detachment' (Eyles 1988, 9). This, however, brings with it the difficulty of negotiation between multiple roles in the field and requires a critically reflexive engagement with one's own positionality, the general research process and inherent power structures (England 1994; Rose 1997; Davies 2008; Bachmann 2011b).

Immersion in the Field

In my case, the process of finding my position in the field was influenced by different factors. I had conducted research in East Africa before and could build on a number of contacts that I had developed during these earlier fieldwork phases (of a total of ten months between 2005 and 2008). In addition I negotiated affiliations with various institutions in Nairobi. Most notable in this context was an affiliation as research associate with the Delegation of the European Union to the Republic of Kenya between January and April 2012. Moreover, I initiated a partnership with the University of Nairobi's Institute of Diplomacy and International Studies (IDIS) – this affiliation eventually also led to a further affiliation with the Kenyan Foreign Ministry's Foreign Service Institute (FSI). Both connections (with the EU Delegation and with IDIS) were instrumental to the research process and facilitated primary data acquisition in various ways. In each case, personal relations with key individuals at both institutions played a significant role. Doing fieldwork and arranging oneself within the research environment and the general situation in the 'field'

> involves hanging around, listening, watching, and otherwise learning the ropes. Establishing and maintaining relationships based on trust and cooperation depend on the deliberate use of commonsense abilities and strategies for gaining rapport and making friends with people within particular situations. These strategies include being open and willing to listen to other people, seeking out common interests, self-disclosures, and establishing common experiences through joint participation. (Jorgensen 1989, 80)

It is often such gatekeepers that, according to Kearns (2000, 114), facilitate 'opportunities to interact with others in the chosen research site' (see also Evans 1988; Enengel et al. 2012). In my case, they not only provided numerous contacts to interview partners and important background information but also served as equally invaluable discussion partners during the fieldwork process for reflection and broader contextualisation. On a very general level, Campbell et al. (2006, 98) define gatekeepers as 'those who provide – directly or indirectly – access to key resources needed to do research, be those resources logistical, human, institutional, or informational'. In their insightful account, Campbell et al. focus on institutions as gatekeepers as opposed to individuals, yet at the same time acknowledged how relations between the researcher and institutional gatekeepers were also mediated through individuals and therefore personal relationships remained 'clearly important in facilitating access through institutional gatekeepers' (Campbell et al. 2006, 114). This was precisely the case for my research as I had affiliations with two institutions, the EU Delegation in Nairobi and IDIS. Both affiliations fulfilled important gatekeeper roles. The former was thereby much more an institutional gatekeeper in a sense that my affiliation with the EU Delegation fulfilled important gatekeeping roles. The affiliation with IDIS on the other hand was rather defined through my personal relation with a key individual. I will now describe the gatekeeping functions of both affiliations in turn.

My Affiliation with the EU Delegation

Initial contact with the EU Delegation in Nairobi was established through an informant I had interviewed in 2008 as part of an earlier project and with whom I had become personally acquainted during my various field work phases in Nairobi. He worked at and facilitated my affiliation with the EU Delegation. Through this affiliation, I was hosted on the premises of the Delegation and offered an office in exchange for sharing my research results and for establishing contacts to (local) academia. Beyond this loosely defined arrangement I had no binding obligations to the EU Delegation.

My office was next to the meeting room at the Delegation. Almost all meetings were held there, including those of the European ambassadors in Nairobi, those of the Political Counsellors (who are normally also the Deputy Ambassadors), the video conferences with Brussels and other meetings. The location of my office immediately next to the meeting room was fortunate for participant observation and for conversations with people waiting to attend a meeting. When the room was still occupied, people waiting to enter often stepped inside my office and started talking to me – sometimes just chatting, sometimes about substantial issues they were just working on. During my first week at the Delegation I tried to negotiate access to key meetings. I approached the responsible EU official at the delegation, Paul,[2] with that request but received a negative reply. The following shows my email exchange with him in this respect:

> *VB*: Do you think it would be possible for me to sit in the meeting tomorrow morning? A large part of my work is trying to understand how the different European diplomatic services operate and how coordination of EU external relations works; such meetings are therefore of great interest to me. I would of course respect any confidentiality issues that might arise.
>
> *Paul*: I am a bit hesitant to include you in this morning's meeting. I would like to discuss the matter with MS [member states] today and seek their views on it. This is important since those meetings are confidential and MS need to feel free to openly express their views on current matters regarding EU-Kenya policies. I hope that you will understand my concern.

Unfortunately, I was also not able to join the meeting or any subsequent ones. Paul had raised my request but it was met with reservation. This exclusion serves as an example of how I remained an outsider during my research process and could not always negotiate the position as an insider I sought to attain (Herod

2 All names are changed for reasons of confidentiality.

1999). Nevertheless, my affiliation with the Delegation provided an invaluable asset because it facilitated access to the Ambassadors, Deputy Ambassadors and Political Counsellors of the member states and, when approaching them, I could refer to my affiliation with the Delegation. I was therefore not a complete stranger to them. Most of them had heard of me before as Paul had presented my request to join their meetings as described above; moreover being affiliated with the Delegation gave me some legitimacy and perceived institutional backing. I was 'no longer treated as a complete "outsider" but as someone who has gained a degree of access to high-level officials both on a professional/research basis but also on a personal basis' (Herod 1999, 324).

My key personal gatekeeper within the EU Delegation was Michael. Michael was roughly my age and one of the first people I met at the Delegation. He also lived just a few houses away from where I lived in Nairobi, so we often commuted together to and from the Delegation. These phases of being together in the car at the beginning and end of a working day were important times in my relation with the gatekeeper. In my research diary, I described this relationship at the time as follows:

> Every day I am stuck in traffic for between 30–60 minutes. Since I am getting a ride with Michael, this leaves us plenty of opportunities to talk about all kinds of political and/or EU-related issues. I am thus in a constant state of researching, however, he also develops into a friend. In no ways do I want to exploit him, I like him and it is nice and enjoyable to be talking to him. We exchange as colleagues, as friends, as politically and EU-interested people. I think he enjoys the conversations with me, too. Still, of course, I also learn things for my research. This problem often appears when conducting research … (RD Note, January 2012)

Indeed, such problematic and ethical dilemmas between personal and professional relations appear frequently when conducting research (Bachmann 2011b). As such they are also indicative of how complicated the research process is and, in particular, how entangled with 'normal' life it is (Friedrichs and Kratochwil 2009). The clear separation between time spent on research and personal time in the field is therefore not possible – and often not even possible when leaving the fieldwork site. Campbell et al. (2006, 117) argue that 'even though we may travel to and from our field sites, we are always "in the field" in the sense that we do not leave this network of social relationships once it has been established, especially our relationships with gatekeepers'. The relationship with gatekeepers is therefore more complex than a simple facilitation of contacts or access from one person to another. My relationship with Michael developed into friendship. However, also professionally, it was not exclusively a process of me extracting information from him – in particular during the hours of commuting and being stuck in Nairobi traffic, I frequently discussed with him recent developments as regards the EEAS, some of the research data obtained or more generally the political situation in

Kenya and the EU. Of course during our conversations I learnt a lot from him about the internal dynamics and the functioning of the Delegation and the EEAS, but it was not restricted to this. Michael also played the role of a colleague with whom I could exchange and discuss my research. He provided feedback and contextualisations to some of the empirical information I obtained. Campbell et al. point to this complexity of researcher-gatekeeper relations and the 'dynamic role of gatekeepers in the many stages of geographical field research' (Campbell et al. 2006, 103); consequently they suggest that 'instead of the "door" model of gatekeepers providing access, we need a "revolving door" model, in which access is always up for negotiation' (ibid., 117).

Such affiliations, however, as in my case with the EU Delegation, also impact on the researcher's perceived neutrality. In particular as regards the cooperation with Kenyan partners, being seen as an affiliate to the EU Delegation puts me in a less neutral light as a researcher. Even though I was not employed by the EU Delegation, nor did I have any obligations to them, having an office in the Delegation building, being European and working for a European university labelled more or less as a Delegation staff member. Even though I normally explained my situation as an independent researcher from Goethe-University Frankfurt who cooperates with the University of Nairobi and is hosted by the EU Delegation to study the processes of establishing the EEAS, I often had the feeling that I was perceived as Delegation staff when interacting with Kenyan informants. For instance, during an interview with an official from the Kenyan Office of the President, he was frequently personifying the EU as me. He was pointing at me and repeatedly said 'YOU' when referring to the EU; when talking about the security situation at the Horn of Africa, for example, he explained how 'EU has a big role to play – *you* are an important actor!' (Author interview, March 2012). This informant perceived of me much more as part of the European external relations' 'operating machine' than as a researcher, even though I explained to him my project and role.

In another case, I had long tried to establish contact with Victor, the head of an important Kenyan NGO. He was a former student of Walter, the professor at the University of Nairobi I cooperated with most closely. When I eventually met with Victor, we started talking about Walter and had a long, lively and interesting but informal exchange that eventually morphed into the interview. I am not absolutely sure if I mentioned my affiliation with the EU Delegation to him – or if he overheard it – however, two days later Victor came to the Delegation as he had a meeting with Paul and Michael and was very surprised to find me there. I noted in my research diary that

> apparently I did not tell him that I was seated at the Delegation. I thought I did, but he was not aware of it. Maybe he didn't fully listen or maybe I just forgot to tell him as I was meeting with a lot of people that day. Anyway, he seemed quite surprised by this and I actually felt a little embarrassed, because in my interviews, I of course try to be independent – which I am. He even said: "I thought you were independent". It is important for my informants to know that I

> am not working for the EU, but that I am researching different EU practices and perceptions. I got the feeling that he felt a bit tricked by me. It just illustrated the importance to be open about your affiliations and backgrounds. (RD Note, February 2012)

I subsequently interacted on several occasions with Victor and while I did not think this incident had spoilt our relationship to any sustainable extent, however, it made me aware of paying better attention to transparency and to communicate more directly my background and affiliations. An affiliation with one institution might produce important insights for studying the internal dynamics of this particular institution, but it might also produce negative effects when trying to cooperate with other informants. It is therefore essential

> to think carefully about the trade-offs involved in working with one gatekeeper instead of another, for example, or about how we manage the gatekeeper relationship. However, we cannot assume that such efforts will have the desired effect or that we can control our image. In spite of a researcher's best efforts to establish her or his independence from a gatekeeper, research subjects will make up their own minds about this relationship based on a number of factors, many of which are beyond the researcher's control. (Campbell et al. 2006, 116)

This naturally also impacts on the researcher's positionality as insider or outsider. Whereas I was perceived as a more or less unwelcome outsider to the EU meetings I tried to join, I was perceived as an insider, someone working for the EU Delegation, by the EU's external cooperation partners, such as Victor. A rigid separation is therefore not possible as the 'stability of the dualism of positionality (the researcher as either 'insider' or 'outsider') is again questioned' (Herod 1999, 324). Depending on the viewpoint, I played both roles in the context of my research (see also Bachmann 2011b). Clearly, this multitude of roles brings significant advantages to the process of gaining information and empirical data – ideally it provides 'an insider's account with an outsider's detachment' (Eyles 1988, 9). Yet, it can also compromise the researcher's perceived independence and therefore be counterproductive for other informants' willingness to cooperate and provide information. These 'trade-offs' need to be considered in each individual case and well thought through and weighed against each other.

My Affiliation with the University of Nairobi's Institute of Diplomacy and International Studies (IDIS)

The second key gatekeeper for my research work was Walter, a well-established International Relations professor at the University of Nairobi's *Institute of Diplomacy and International Studies* (IDIS). Many of Walter's former students now occupy key positions in the Kenyan government or the civil society sector. I initially approached Walter to inquire about his interest in research cooperation in the context of the EuroGaps-project, which would involve thematic research cooperation as well as research and student exchanges. He accepted my proposition and the exchange began formally with my first trip to Nairobi in November 2010. IDIS is a key international relations education and research institute in Kenya; between 1982 and 2007 it trained most of the Kenyan diplomats and foreign policy officials. Now it also continues to produce international relations experts for governmental and non-governmental organisations. Walter was therefore extremely well connected and possessed an extensive network of connections to Kenyan elites in the area of international relations both in the Kenyan government and beyond. Moreover, I had the impression that, due to his position as a professor and mentor, most of his former students regarded him very highly and felt obliged to positively respond to his requests. Gaining his trust and will to cooperation was an invaluable asset to access key informants on the Kenyan side. In a similar situation as regards her fieldwork in Benin, Jennifer Mandel (2003, 203) recounts how

> [t]he professor with whom I worked in Benin was, indeed, very helpful in obtaining the necessary authorisations from the national and local government. He was fairly well known and respected as a researcher and, before I left Benin, he was appointed as a minister in the government responsible for the development of small and medium enterprises. Armed with a letter from him, I had no trouble.

My interest was neither in enterprises nor did I require authorisations, however, the affiliation with IDIS, the connection to Walter and his facilitation of contacts opened many doors for me. Most instrumental, for instance, was his assistance for the trip I had planned to undertake to Juba, South Sudan. I had long envisioned a trip to South Sudan, mostly out of curiosity and interest in this newly independent country, however, also out of a research interest in the difference between how collective European external relations function in a highly malleable environment of emerging state structures in comparison to a relatively established spaces of diplomatic conduct in Nairobi (see Bachmann 2013d). I had talked to Walter about my planned trip on several occasions and he mentioned that he would like to introduce me to the South Sudanese Deputy Ambassador to Nairobi, who was also one of his former students. The evening before my planned departure to Juba, Walter introduced me to him. On this encounter I noted the following in my research diary:

> After I talked for some time to the Ambassador, he asked me if I already had my visa. Of course I didn't as I read on the AA-website [AA = Auswärtiges Amt (the German Foreign Ministry)] that it is possible to get the visa once you arrive at Juba airport. So I didn't bother organizing anything in advance. Well, according to the Ambassador, this is only possibly when you come into the country [South Sudan] from a place that does not issue visa. But of course, the South Sudanese Embassy in Kenya does! It was the first one they opened. So I *has* to get the visa

> before the trip, otherwise they would not even let me enter the plane. It was about 17:30 when I was informed about this, and my plane left at 7:30 in the morning. Shida Kubwa! [big problem] Well, Walter told the Ambassador to arrange things so that I could get the visa that night. And he did! The Ambassador just replied with a smile: "well he is my boss (professor), so I have to do what he says!" When I told Walter thanks for his help, he just responded: "No problem. That's part of our cooperation!" What happened then was that the Ambassador called one of the staff at the Embassy to get back to the Embassy and issue me a visa. So, at 21:00 last night, I walked into the brand new, and then deserted and wide open, South Sudanese Embassy and was issued a visa for "Officials". (RD Note, February 2012)

Of course these proceedings raise a number of ethical questions on the research process. However, it is also exemplary for many occasions where the affiliation with IDIS, or rather the personal relations with Walter, decisively facilitated the research process. In other instances, Walter used his connections to facilitate access to key officials and managers of non-governmental organisations, both for conducting interviews as well as for participation in panel discussions and focus groups.

Methodological Remarks

Data Acquisition in the Networked Places of EU Diplomacy in Nairobi

> [T]riangulation demands creativity from its user – ingenuity in collecting data and insightful interpretation of data (Jick 1979, 610)

This study pursues a critical geopolitical research approach that is sensitive to both the places and social dynamics through which European diplomacy is enacted in Nairobi as well as the local and wider geopolitical contextualities within which such enactment takes place. Through affiliations with the EU Delegation and the *Institute of Diplomacy and International Studies* I had important gatekeepers that facilitated my access to informants and allowed for deeply 'immersing in the field'.

Inspired by Friedrich and Kratochwil's (2009) call for pragmatism in social science research and the need to find a compromise between scientific rigour and knowledge production in everyday social practice (ibid., 714), I employed multiple methods of data acquisition and analysis. During fieldwork, I used three different methods for obtaining data. Such methodological triangulation has long been debated for its pros and cons in the social sciences (Vidich and Shapiro 1955; Webb et al. 1966; Hemming 2008; Taylor 2013). By bringing together different research practices, Yeung (2003, 456) suggests a process-based methodological framework that is 'explicitly iterative and dynamic, rather than linear and stage-linked'. It is through such an iterative and also explorative approach, and in the context of Friedrichs and Kratochwil's (2009) call for pragmatism in social science research, that my methodological triangulation developed from an initial focus on semi-structured interviews to eventually also include focus group discussions and research notes based on participant observation – both emerging out of my affiliations with the EU Delegation and IDIS. Nevertheless, interviews remained the key method of data acquisition.

The interviews were roughly structured into three parts. The first was a more general part in which I inquired about the processes of the construction of the EEAS and the 'technical' details of the new service. I asked informants about how they had first heard about the EEAS and to describe the service in general terms and how it impacted on their work (in the case of EU and member state official) or their cooperation with the EU (in the case of the Kenyan cooperation partners). I aimed to obtain the informants' *"objective" description* of the EEAS. The second part of the interviews sought out the informants' *subjective opinion* of the new service. I asked what they thought about the EEAS, to evaluate the EEAS' early phase and their work in or cooperation with the EEAS, and if they regarded it as 'useful' for the EU to build a collective diplomatic service and to conduct foreign policy collectively. I also asked informants about what they thought the EEAS *should* do to facilitate their work or the cooperation with the EU. In the last part of the interviews, I moved the discussion away from the EEAS and inquired about informants' opinions and perceptions of the EU in general, both as regards its role towards Kenya and as a major global actor. I asked informants about which role, in their opinion, the EU *should* play in Kenya (or the wider East African region) and in the world more generally. In this context, I usually ended the interviews by asking informants about the three most positive and most negative aspects they associated with the EU in general.

The three parts were of course not clearly separable, yet they provided a rough structure for the conduct of the interviews and, thus, for inquiring into the processes and practices of how the EEAS is *positioned* and *perceived* amongst those that enact it externally and those that cooperate with it. Given the site of the study in Nairobi, the interviews were conducted in the broader context of Kenyan-EU relations, yet the topics discussed were much broader. Most of the informants had substantial diplomatic and foreign political experience. When interviewing EU and member state officials, I specifically asked to situate their responses in their wider experience of the various postings they had held during their careers, either in national diplomatic services or in the EU system. Equally, all of the Kenyan governmental and non-governmental informants were experienced diplomats or functionaries, having occupied a range of diverse posts in the public or private sector. I emphasised that I was interested in their opinion on situating the cooperation with the EU in the wider contexts of their experiences of cooperating with diverse international partners throughout their careers. As such, the discussions generally covered issues beyond the immediate Kenya-EU relations. Through their broad experience, informants were able to provide insights into wider processes of EU external relations, much beyond the EU's role in Kenya, as well as to situate the EU in the wider context of international actors – both in Kenya and amongst major geopolitical players.

In total I conducted 58 interviews with 56 informants; 11 of these informants were EU diplomats or officials, 25 were diplomats or officials from EU member states and 20 were diplomats or officials from the EU's cooperation partners in Kenya (Kenyan government and non-governmental or civil-society organisations). In addition, I conducted three focus group discussions lasting roughly two hours each that involved a total of 75 participants. One of these focus groups was a high profile panel discussion amongst 14 senior foreign political experts, including the Deputy Ambassadors of the EU, France and Germany as well as the former Kenyan Ambassador to the EU, the head of one of the most prominent Kenyan NGOs and several academics. The other two focus group discussions were part of two classes I taught to a total of 61 students at IDIS. The classes were held in the evening hours as they were designed for professional students in pursuit of an additional advanced degree at IDIS. Most of those students were highly qualified African professionals with significant foreign political expertise, aged between

25 and 60 and working for government institutions, embassies, or international organisations in Nairobi. Having access to these foreign political professionals (from different angles and national backgrounds) afforded me the possibility to include their perceptions of the EU and its geopolitical role into my research in the form of transcripts of the discussions and through a short questionnaire handed out to them before class. The questionnaire covered some of the questions asked in the interviews and subsequently discussed in the focus groups. As such, and due to its small sample size, it is not to be treated as a representative survey. Rather, I used my exposure to this group of foreign policy professionals to ask about their thoughts on and experiences of cooperating with the EU. Instead of employing the questionnaire as a statistical, representative method, I thus used it as a complementary data source for inquiring – in written form, yet in a similarly qualitative manner as the interviews – about informants' subjective opinions and perceptions of the EU.

In total, the number of informants questioned for this research is 124: 56 in interviews, 14 in a high profile focus group discussion (of which 7 were also interviewed) and 61 in two more focus groups and the preceding questionnaire. The mismatch of having 25 interviews with diplomats from member states and only 11 each with EU and Kenyan government officials results from the availability of individuals suitable for my research. During the course of my research, 19 EU member states had embassies in Nairobi, thus offering a higher number of possible interview partners than could be found in the 'singular' institutional structures of both the EU and the Kenyan government. In addition, my affiliations with the EU Delegation and IDIS provided ample opportunity for participant observation. The information obtained through interviews and the focus groups, was therefore supplemented by research notes I took during the various fieldwork periods. Throughout the text, quotes from interviews are presented in the form 'Author interview, *Month Year*'; the identity of the informant is kept anonymous, yet I indicate her/his affiliation. Excerpts from the focus group discussions and the preceding questionnaire as well as from the research diary are presented as 'FG, *Month Year*', 'FG questionnaire, *Month Year*' and 'RD notes, *Month Year*' respectively. Table MR.1 (below) provides an overview of informants and the primary data sources this book is based on.

I proceeded with the primary data through a qualitative analysis using the programme NVivo9. For coding and developing categories, I used a mix of previously identified thematic directions and of inductive analysis during and after data acquisition. Such inductive analysis, according to Patton 'means that the patterns, themes, and categories of analysis come from the data; they emerge out of the data rather than being imposed on them prior to data collection and analysis' (Patton 1980, 306). I developed the exact categories used for coding throughout the research process, yet certain broad orientations existed at the beginning of the research process. Drawing on Blumer's (1954) pioneering work on sensitizing concepts, Bowen (2006, 14) argues that researchers 'might use sensitizing concepts simply to lay the foundation for the analysis of research

Table MR.1 Overview of informants and primary data sources

	European officials and diplomats		Cooperation partners		No. of informants
	EU	Member states	Kenyan Government	NGO/CSO	
Interviews (58 conducted)	11	25	11	9	**56** (755 codings)
	36		**20**		
Focus groups (3 discussion rounds)	**4** **0** **0**		**10** **41** **20**		**75*** (96 codings)
Questionnaire survey	**0**		**61**		**61** (505 codings)
Research notes	55 entries in the research diary				(88 codings)
Total	**36**		**88**		**124** (1444 codings)

* Of the 75 participants in the focus group discussions, 61 also participated in the survey and 7 were interviewed.

data. Researchers might also use sensitizing concepts in examining substantive codes with a view to developing thematic categories from the data'. In this light, the broad orientations I set out to inquire into were then modified and refined during the field work and data analysing processes; also some were added, others removed. Based on these broad orientations, a continuous grounded theoretical reflection during the research process helped to build the categories in which obtained data was thematically structured on three levels of specificity. Interviews were the methodological centrepiece of this research, yet also the information and experience obtained through focus groups, participant observation and more pragmatic, informal research methods that was recorded in the research diary, was an invaluable source of information that allowed me to make sense and better situate the interview data. Based on such methodological triangulation, the second part of this book now presents the analysis of the making of EU diplomacy in Kenya.

PART II
The Service in Action – Establishing Collective Diplomacy in Nairobi

Chapter 5
Positioning Collective Europe Abroad: Operating EU Diplomacy

> The EU is a little bit like a marriage. It is often difficult, it has its problems, it's not perfect at all, sometimes it makes you tired and you're thinking of dropping it. But at the end of the day it is such a valuable and good thing always worth fighting for. (Author interview with member state diplomat, February 2012)

Decades of seeking to regulate the spaces of interaction have left the EU with a particular mindset of operation and EU agencies with a particular *Esprit de Corps* (Benson-Rea and Shore 2012). In Chapter 1, I have outlined how the establishment of the Commission's early external delegations has been accompanied by a struggle of self-definition of external staff to see themselves as technical implementers or as diplomats (Carta 2012). This struggle continues as part of the processes of establishing the European External Action Service. The EU's relations with Kenya, and more generally with countries where development cooperation has traditionally played a large role, have long been characterised by long-term technical arrangements of development cooperation and trade policy. Part of these largely technical processes of implementing development policy was a standardisation of the spaces of interaction through which trade and development is implemented. This, of course, has to be seen in the wider picture of the EU's strive for regulating interaction spaces – both internally as well as with its external partners (see Chapter 3). Before the Lisbon Treaty, these dominant and largely regulated policy fields of Africa-EU relations, i.e. development and trade, were also the key areas of Commission competence and where, consequently, collective policy conduct was most pronounced. Foreign policy and diplomacy remained in the hands of the member states. The collective aspect of Africa-EU relations was therefore perceived as largely 'apolitical', i.e. less influenced by changes in political relations between the partners, as they might occur through a change of government for example. As a result, in the words of a World Bank official, cooperation with the European Commission 'can be more stable and objective than that with its member states because it is not so much affected by political fluctuation. The Commission's approach is more long-term oriented and issue-based. For us, it is easier to work with the EU as it is apolitical, just like the Bank' (Author interview, February 2008). Similarly, a Canadian official observed that the EC is 'not so susceptible to political issues and possible bilateral problems. It is more consistent, the EC can ride stormier seas' (Author interview, November 2008).

The EU, as the collective European polity, was thus seen largely as a depersonalised and technocratic/bureaucratic agent in the areas of development and trade. The changes provided by the Lisbon Treaty, however, sought to establish a more political role for the EU through the EEAS. In an early engagement on the future of the EEAS, Caterina Carta (2012, 156–9) suggests that the EEAS will have to establish its position on a 'political continuum' between being a broker and being an entrepreneur. The former largely refers to a broker/coordinator for the foreign policies of the member states and towards the Kenyan government, while the latter describes a more active role as a largely independent diplomatic agent. It is such processes of negotiating a collective diplomatic role between the different agents of European external relations in Nairobi that will be addressed in this chapter. Based on empirical material from interviews with diplomats as well as from participant observation through my affiliation with the EU Delegation in Nairobi, I describe how political and diplomatic relations are coordinated and how 'Europe' positions itself collectively. I thereby focus on the first step of diplomacy as outlined in Chapter 1 and the practices and processes of *Europeanisation abroad* (see Chapter 3), i.e. the coordination and conduct mechanism of policy between the Delegation, EU actors in Brussels and the member states. It is comprised of three sections plus a chapter conclusion. The first section provides an overview of the procedural changes and the staffing processes in the early phase of the EEAS. This is followed by a section on the operationalisation of collective diplomacy, explaining how the service operates in the specific context of Nairobi and struggles with adding a political role to the previous role of the Commission's Delegation as a largely technical development implementer. In a third section, the relation and competencies between the different actors of European diplomacy is discussed.

Establishing the EEAS

The EEAS was not created from nothing. Whilst it did entail new diplomatic structures for EU agencies, a major part was also aligning the established diplomatic structures of the EU's member states and the external relation instruments of the Commission and the Council. The special situation of the establishment of the EEAS needs to be taken into account in this context (see Chapter 1 and 2). The Lisbon Treaty had been waiting for more than two years before its final ratification in the Irish referendum on 02 October 2009 and its eventual entry into force on 01 December. The contents of the Treaty and the changes that were about to come for the European Union and its external relations (in particular the establishment of the EEAS) had been known for a long time, but they had also been suspended due to the uncertainty about the ratification of the Treaty. Once the Irish 'Yes' vote allowed for final ratification, however, time was pressing. The Treaty entered into force less than two month after the vote; the changes foreseen by the Treaty then required rapid preparation for it to be implemented. Whereas European officials in

Brussels seemed reasonably well prepared, the changes for the conduct of external relations seemed to catch European personnel abroad, in particular member state diplomats, short and without adequate briefings. It was a peculiar situation, since the diplomatic services had several years to prepare for changes that might eventually come (or not), but once it was certain, time was lacking and hardly any information on the practicalities and details was available. A member state diplomat recounted their dilemma:

> If you have been working in diplomacy and the foreign ministry for a long time you just know about Lisbon and the entire process […] in many ways we were directly involved in the preparation for final implementation of the Treaty. Ratification then came so quickly, in a way much quicker than most expected. Then the big question was: HOW? How is this now being put into practice? (Author interview, March 2012)

Most of the information provided to member state diplomats came through the established diplomatic channels of their foreign ministries, not from the Delegations – which were equally caught off guard by the rapidity with which the ratified treaty was being put into practice at the turn of 2009 to 2010. Whilst the EEAS was formally established with a preparation phase of one year on 1 December 2010, some of the Delegations, including the Nairobi one, were transformed from Commission Delegations to the EU Delegation at the beginning of 2010 (see Chapter 1). This included taking over the duties and responsibilities of the troika of past, current and future presidencies of the European Council, such as setting the agenda, organisation and hosting all regular meetings. Neither EU external relations personnel nor member state diplomats were prepared for this rapid change. An EU official in Brussels recounted how

> for the delegations it was a massive challenge as they only had a little more than one week to fill an entire range of new obligations, such as basically running the presidency They did not get any additional funds, personnel or facilities. There was quite a bit of frustration going on. (Author interview, July 2012)

The frustration caused by a failure of communication as regards the urgency of implementing the Lisbon Treaty after years of political stalemate, was widely felt. It was thereby not a lack of information about the broader processes around the Lisbon Treaty that diplomats criticise; they were well aware of the wider process. Rather it was a lack of information of what this entailed for the practicalities of conducting foreign policy and diplomacy, both with regards to member states, but also with respect to an emerging *collective* diplomacy. Diplomats from the member states were simply 'expected to know what is going on' (Author interview, November 2012). The situation was slightly different with respect to the training of EU external relations staff, hired specifically as EEAS diplomats. Though limited in time and scope, new EEAS staff underwent

> a week-long training in Brussels for Heads of Political Sections. Ashton was not there, but her Secretary-General Pierre Vimont. There was a briefing on the expectations. Not really any guidelines on what to do. Guidelines on how to write political reports to Brussels. Guidelines on how to write local statements. The rest needs to be found out oneself. (Author interview with EU diplomat, January 2012)

As the informant here lays it out, the training was focused on guidelines for reporting rather than on practical implementation and how to enable the new diplomatic actor, and much remained to be 'found out oneself'. Ashton's lack of experience in diplomatic matters or in terms of leadership certainly did not help facilitate a smooth start for the diplomats involved in building up the new service. However, given that the Nairobi Delegation had to operationalise the changes provided for in the Lisbon Treaty eleven months before the official start of the EEAS in Brussels, there was a certain grace period in which to negotiate competences. During that phase, in May 2010, Ashton made her first visit to Africa, though this seemed to have left many blanks as regards the orientation and operationalisation of the service. An EU official who was present at Ashton's visit to Nairobi plainly stated that she 'kept a low profile and has no clue what's going on. She is a good listener but not more […] Well, she was chosen because the European governments wanted a low profile figure' (Author interview, November 2010). Two years later the same informant repeated a similar criticism of Ashton's lack of experience and of her leadership style, criticising how the

> communication from top-down has been extremely poor. She has no foreign political experience or knowledge. When she first came to office she was asked to do a briefing on […] She responded that she could not do that as she did not have her service operating yet, but she had more than 200 RELEX staff and the people from the council to work with. Still she couldn't do it. (Author interview, July 2012)

These statements are indicative of the wider sentiments expressed against Ashton during her first months in office both by observers and partners of the EEAS as well as within the service itself (see also Howorth 2011). During the first two years, the EEAS' immediate effects on the coordination between member states and delegations were also described as rather limited, in particular with respect to the collective *political* role of the EU. Most visible was the taking over of the roles of the rotating presidency by the Nairobi Delegation as opposed to the troika of past, current and future presidency as before. Member state diplomats described the effects in rather similar ways:

> The official launch of the EEAS, did not have a very big influence, except for when your country holds the presidency. Then of course it had a big influence because the EU (delegations) took over a lot of the functions you were supposed

> to be taking care of before. Maybe it also changed in a way that the "collective" bit is now a bit more consistent since the agenda is now set always by the delegations. (Author interview, November 2012)

> As regards the bilateral relations with Kenya not much has changed. But you need to consider that we are a small country and basically have no bilateral interests here. Before the launch if the EEAS it was the country of the rotating presidency who made statements on behalf of the EU, now it is the Delegation. For us, with almost no political interests here, it does not really matter who speaks on behalf of the EU.
>
> […]
>
> Nonetheless, for the smaller member states it facilitated things as they had insufficient manpower to run the show. (Author interview, January 2012)

Beyond this organisational shift of replacing the troika through a more constant organisation and of agenda setting at the Delegation, the immediate effects of the Lisbon Treaty on the conduct of collective diplomacy remained limited in the first two years. In a somewhat sarcastic but also indicative way, a member state diplomat remarked how 'the most notable thing of the EEAS' establishment was the change in the email-addresses of the people to deal with. Other than that not much changed' (Author interview, January 2012).

Staffing the Service

Once the service had officially been established, it needed to recruit its staff. With a total of 3,400 staff (as of July 2013), this was a major undertaking. Despite the decision to establish the EEAS as an agency independent of the Commission, the practicalities often required Commission staff to take over functions that were assigned to the EEAS through the Lisbon Treaty. Heads of Delegation (HoD) who started their post before the ratification of the Lisbon Treaty were under a Commission contract. With the establishment of the EEAS they were no longer heading Commission delegations but EU delegations; for some, the time given to adapt to and implement the changes was one year (from the start of the Lisbon Treaty in December 2009 to the official launch of the EEAS in December 2010), for others it was only one month (until January 2010, see Chapter 1). In particular the newly ascribed political role turned out to be problematic for the delegations since they were supposed to be taking over coordinating functions (such as setting the agenda, preparing and hosting the meetings, drafting communiqués, etc.) but did not have 'political' staff. For other policy fields, such as development or trade relations, the coordinating role was easier to fill since delegations could rely on present Commission staff to do so. Especially in its early days, the EEAS could not have fulfilled its mandate abroad without the assistance of Commission staff and member states embassies. The dividing lines between the mandates of Commission and EEAS thereby remained blurred for the transition period between the time when the EEAS was supposed to take over certain functions until it had the staff to do so. Depending on the delegation, this transition period ranged from several months to several years.

Even though the staffing process officially followed clearly defined procedures, as in any institution, much depended on personal contacts and, as in any EU institution, also on national interests and certain quotas to be filled. It is worth citing here at length from an interview with an EU official who gave an example of how, in particular for more senior posts, the juggling of interests and quotas played a key role. The informant explained how

> [s]ometimes certain member states insist on certain posts, the Dutch wanted the Head of Mission [also Head of Delegation] in South Africa. Spain the one in Namibia – for fishery issues. Now there is an Italo-Chinese connection handling affairs in Uganda. There is oil and an Italian company had an interest in Ugandan oil. There is a lot of carving up of areas of interest around the globe through the member states, they use the EEAS to pursue these kind of geopolitical interests. The member states can still determine a lot of the outcomes.
>
> Question VB: *How, then, is the procedure to find a Head of Mission?*
>
> This is a good question, I will tell you a story that I heard from a good colleague on how the Head of Mission for the Delegation in *Fictisia*[1] was determined. There was a list of 5 people and then a Spanish guy (a senior national diplomat) asked why there was not a Spanish person on the list. The person from the search committee responded that the list of 5 people was developed based on the best applications. However, after the Spanish intervention, the list was rearranged, some rankings and marks were changed and all of a sudden the Spanish guy that was recommended by his national colleague was on the list. Take a wild guess who is now Head of Mission in *Fictisia*? This conversation happened over lunch. It is half gossip, but I think the story is roughly true. It seems as if for developing short lists, objective criteria are used and proper rankings developed, but once that short list exists, there is a lot of interferences with quotas and preferences that nobody really knows how it is done. The member states put a lot of pressure on the assignments. France, for instance, issued diplomatic documents for French people who were not part of the French diplomatic service, so that they were eligible for applying to EEAS posts. There is an entire culture of this system that partly came from the Commission. The Cabinet has a lot of influence and say in this; *Kaminski*[2] made an art out of it. My colleague was on travel with him once and recollected how he was permanently on the

1 Country name replaced by a fictious name for confidentiality reasons.
2 Name modified.

> phone and just overheard him saying things like: "No I don't want a French in that position, I want a Greek" and so on. It's based on juggling and rewards. (Author interview, 18 July 2012)

This complicated process of 'national and institutional horse-trading' (Kuus 2012, 9) is an essential feature of the functioning of the European Union and the processes governing how positions, in particular senior positions, are filled (see Howorth 2011; also Chapter 1). More generally, the personal preferences, connections and networks of individuals in charge are key factors for influencing the functionality and agency of any international organisation. I have elsewhere argued for the need to conceive of international organisations (including the EU) as sites of social interaction, whose agency is shaped – though not exclusively – by the internal and external social relations of the people that make up the organisation (Bachmann 2013c). More than two decades ago Nohria had already pointed to the 'informal or "emergent" relationships' contained in and constitutive of international organisations and argued that 'identifying and analyzing these "hidden" networks can be of great significance in understanding organisations' (Nohria 1992, 5). More recently, a number of scholars in critical geopolitics have made significant contributions towards studying the internal dynamics of organisation. Müller, for instance, has taken up Nohria's suggestion for identifying and analyzing the hidden networks in organisations in his insightful ANT-inspired call to open the black box of the organisation and 'render visible the arrangements contained within them' (Müller 2012, 386). Kuus, too, has traced the 'social alchemy [and] social hierarchy' (2010b, 386) in the European quarter in Brussels in the context of enacting bureaucratic expertise (2011a) and geopolitical knowledge generation (2014).

Such 'hidden networks', 'arrangements' and 'social alchemy' of the individuals staffing the EEAS were frequently alluded to during the interviews and observable during lunches, coffee breaks, informal chats as well as during formal and informal functions I was invited to. In particular for the process of linking the EEAS with the diplomatic missions of the member states, for gaining trust, these social networks play a crucial role. An EU official explained how former national diplomats that moved to the EEAS 'probably maintain their personal relationships with the former colleagues and exchange information informally, but this is not part of the set-up, officially they break the ties with their national ministry' (Author interview, November 2012). As part of national staff in EU institutions, Geuijen et al. (2008, 117) recount how Dutch diplomats describe their roles using words such as 'feeler', 'resonance box', 'ambassador', 'antenna', 'brainstorming partner' but also 'missionary' and 'infiltrator'. Similar sentiments frequently appeared in my interviews. For instance, another EU official reiterated a similar understanding of this personally enacted relationship between member states and the EEAS, recounting how *Clive Sailer*,[3] a senior EEAS official,

3 Name modified.

> receives information through a variety of channels; some through the EU desk offices, but also through his old UK connections. The old networks are still ongoing. He has 2 cellphones. One from the EU and one from the UK Foreign Office. He said "they let me keep the old phone, I can encrypt the UK message with it". Well, of course they let him keep it … (Author interview, July 2012)

Such 'dilemmas' are of course only possible because of the novelty of the EEAS as 'the first diplomatic corps anywhere that is not in the service of a nation-state' (Kuus 2014, 5), creating the unique opportunity for diplomats to move from one diplomatic service (their national one) to another (the EEAS). Being in both the EEAS and national diplomatic loops as described above for *Sailer* is, of course, not official practice, yet the double-identity of EEAS diplomats functions as an important factor for building mutual trust between the new institution and its closest allies, the national diplomatic services of the member states. Yet, in particular at the beginning, the staffing of the EEAS caused considerable frustration both within the service as well as amongst the member states. Positions needed to be filled with people with diplomatic expertise – an expertise that was not present in the Commission and in the working culture of Commission agencies, including the Delegations, before. Staffing the EEAS in equal parts with personnel taken from member states, Commission and Council was the intended approach to deal with this problem, as an EU diplomat explains:

> Well, recruiting 1/3 of EEAS staff from the member states was initially intended as a means to get people with diplomatic experience and expertise in the EEAS. Commission people did not have that as they did not play a political or diplomatic role. They are not diplomats, they are technical implementers who administer development cooperation and development funds. The EEAS needed diplomatic expertise, the Commission did not have that; it had to come from the member states. [Also] for the way of thinking of the member states it might simply create more confidence if this new institution is staffed with their 'own' people. It is very important for the EEAS to gain confidence amongst the member states and the partners. You know it is a new institution, nobody knows how it really works, what it is capable of doing and if it can be trusted. Having it staffed with people from your own diplomatic service, people you know you can trust, creates a certain level of confidence. (Author interview, November 2012)

The social networks hinted at above thereby not only function as a working mechanism within the EEAS, but also as a facilitation for the EEAS to gain confidence amongst the member states due to the continuation of many personal relationships in member states' diplomatic services. However, the process did not go uncontested. Due to a shortage of staff and the urgency with which the EEAS needed to be staffed, a large number of Commission staff was 'converted' into political staff for the EEAS without adequate diplomatic experience. An EU official recounted how some of these new 'political offices were quite junior' yet 'they often interpreted their role as being the No. 2 in the Delegation [Deputy Head of Mission], but this was only officially the case for Kenya and South Africa […] This caused a lot of frustration for the more senior Commission staff in the delegations' (Author interview, July 2012). Not only senior Commission staff were frustrated with the hasty staffing of the EEAS and the new role it was supposed to play – including a (perceived) infringement on their previous functions and competencies; also member state diplomats in Nairobi lamented how 'the competencies and the experiences of the member states' foreign services were not at all taken into account' and how the EEAS 'was introduced too quickly and in a rush that did not help anybody. Its level of incompetence is not as high as in the UN, but still very high' (Author interview, November 2012). A large part of that informant's frustration resulted from what he considered a 'very inefficient use of money' in the EEAS and the institution's exploitation by some of its staff as a tool for 'advancing their careers' without any 'real commitment' to it. Indeed much of the frustration amongst both EU and member state officials was due to the rushed set-up and staffing of the EEAS. However, also the undefined relation between different Commission agencies, the EEAS and its political/diplomatic role and its relation to the diplomatic services of the member states has contributed to the confusion and dissatisfaction in the EEAS' early days. I will treat these relations in the following.

Operationalising the Conduct of Collective Diplomacy

> In the absence of clear references to the cornerstones of international identity, European diplomats engineer diplomacy by making sense of the entity they represent while they represent it. (Carta 2012, 4)

> There has been no dissemination on what will happen when the EEAS starts, it was just there. It is learning by doing. You follow the steps [of European integration] all along and then when you start working on it, you just work your way into it. Only by doing it you learn and understand how to act within it. (Author interview with member state diplomat, February 2012)

When establishing the functionality of the EU's diplomatic mission, a considerable portion of improvisatory skill was required from the individuals involved. It is in the nature of such heterogeneous and composite institutions (in terms of nationalities, ideologies and previous institutional socialisation), dependent on the cooperation of other institutions (Commission, other EU agencies and national diplomatic services of the member states), that different opinions about its orientation exist and are fought over. It is also in the nature of such an institution that the heterogeneous and composite staff that constitute and cooperate with this institution have different feelings about the commitment to this institution. EU diplomats seemed generally highly committed to the establishment of a more collective diplomacy – this is unsurprising though, since it provides them with a job (see also Juncos and Pomorska 2013a). The commitment amongst member state diplomats, on the other hand, was more ambivalent. The following quotes from a rather sceptical member state diplomat illustrates this ambivalence: 'In terms of coordination amongst EU partners, there is nothing new, no better coordination than before, maybe even the opposite'. The EEAS, he continued to argue 'was a political decision made in Brussels, the first year was *very* inefficient, the second year was inefficient. But I am a diplomat I am a mercenary of my government, so I implement the political decisions they make' (Author interview, November 2012). However, the same diplomat also acknowledged how 'through the larger EU vehicle, we can now knock on more doors and approach easier more contacts, it has facilitated our access to contacts' (Author interview, November 2012).

A major problem appears to have been uncertainty and vagueness regarding the operationalisation of the service and the lack of leadership in relation to its political orientation. A member state diplomat explained how it was a big a problem 'that many aspects of the Lisbon Treaty were done by people that had no idea about how it should function *in practice*'. To some extent, that lack of leadership on part of the EU was compensated by 'local practicabilities that makes things work' regardless of the conformity to the Lisbon Treaty (Author interview, January 2012). Here too, the social networks and interactions amongst EU and national diplomats in third countries facilitates such 'local practicabilities' the informants hinted at. Nevertheless, the lack of a functioning 'command structure' (Author interview with EU official, July 2012) and the confusion over competencies severely impeded on the functionality of the service.

Prolonged Decision-making and 29 Principals

One of the problems frequently identified was the inertia and slowness with respect to decision-making and releasing collective statements. However, such delay is inherent in the EEAS' composite nature. It has to find a position acceptable to *all* member states *and* the Commission (see discussion on two-step diplomacy in Chapter 1). In contrast to national diplomatic services, the EEAS has to follow not one but 29 principals, it does not represents the position of one single national foreign ministry, but of 28 plus the Commission. Before a political statement can be issued, it therefore has to be negotiated and agreed upon with all the agents represented by the EEAS – it is thus the instrument not of the Commission's single external relations, but of the *collective* external relations of the European Union. Before the actual diplomacy towards third countries (i.e. Kenya in the case of this study), a long process of negotiated diplomacy has already occurred between the EEAS and the member states' diplomatic missions. This need for a concerted statement is often problematic when quick diplomatic reactions are necessary. A national diplomat, for instance, explained that 'if you are on the ground you need to be more flexible. This is of course difficult for such a multifaceted organisation like the EU. The Delegation can just not be so flexible like an embassy'. A real problem for decision-making, the diplomat continued, however, is 'when others always have to go back to their headquarters before they make a decision. We have the mandate to make decisions here, if others don't have that, it drags on and takes a very long time' (Author interview, March 2012).

The lack of leadership on part of the EU is thereby seen as complicating and prolonging the process of *Europeanisation abroad*, i.e. the conduct of the first step of diplomacy between the different European actors in Nairobi (see Chapter 3). During interviews in 2012, several member state diplomats voiced dissatisfaction with the vague role of the Delegation in terms of providing a broad orientation for collective external relations, in particular in the area of development policy. As part of an interview with a national diplomat, for instance, I asked 'Would you be happy with having your name under an EU statement without having been consulted before?' The member state diplomat responded:

> Yes, I would be ok with that if I knew what to expect from the Delegation, but we don't as they never voice an opinion on their own. I would have been fine with it if Erik [van der Linden, the previous Head of Delegation] would still be around. Now we just don't know where the EU stands.
>
> […]
>
> We also do not have to consult [our capital] before we make statements referring to Kenya. Whatever concerns Kenya is in the hands and the authority of the embassies here. This is the same for all embassies. We have that freedom. If someone says: "we need to consult with our capital", it means that they did not engage with the issue yet. We want to hear what the EU has to say. After all it is the biggest donor in Kenya! (Author interview, February 2012)

More generally, during the interviews with EU and national diplomats in Nairobi, I asked informants to describe the process of how joint statements are developed as part of the EU's collective diplomacy. The lack of political orientation and guidance alluded to in the quote above was one aspect that appeared frequently. On the one hand, it was often lamented by smaller member states who wished for a stronger collective role that they could subscribe to. On the other hand, larger member states tended to encounter it with a sarcastic complacency reflecting their expectation of the EEAS' inability to make pronounced and quick statements with political impact. Even though hardly outspoken, a certain feeling of the EEAS' redundancy was looming over the attitude of diplomats from larger member states – seemingly only waiting for the Delegation's reluctance or hesitation on certain topics to readily step in to take the lead. One national diplomat, for instance, explained how 'the British are often not willing to share and cooperate with their other European partners. But they are so well manned and connected that in many ways the rest of the EU is just following the Brits here' (Author interview, January 2012). The British role is clearly a special one in Kenya, not only because it is one of the largest member states of the EU but even more so due to its colonial past in Kenya and the size of its embassy in Nairobi of more than 300 staff. Another member state diplomat described the process of developing joint statements as follows: ‘the delegation usually makes first drafts of political statements, then those are circulated and discussed. *If* there is disagreement, it is often the big ones who disagree. Then the delegation rephrases and makes a second draft and the process starts again’ (Author interview, March 2012). Similarly, a colleague from another national embassy explains how at the start of the process the Delegation ‘spreads some basic information, but the fine-tuning is often done by the member states. They [the Delegation] are intellectually too slow and just don’t have the experience and contacts yet. It needs time to grow and develop these skills and procedures’ (Author interview, January 2012). Developing such ‘skills and procedures’ is an essential part of the process of establishing the spaces of interaction for *Europeanisation abroad* and for *positioning* the EU externally.

Regulation Versus Personalisation in External Relations

In addition to the lengthy process of developing an internally consented position before collective external articulation, national diplomats identified the EU’s ‘obsession’ with following specific, regulated procedures and processes as a key reason for the slowness of collective action. Various member state diplomats pointed to the ‘procedures we need to adhere to’ (Author interview, March 2012) or outlined how ‘in many ways the EU cannot really define what content it wants. It is caught in following procedures’ (Author interview, January 2012). What aggravated the EU’s bureaucratising zeal was the corruption scandal around the Commission headed by Jacques Santer in 1999. The scandal resulted in a further technocratisation and standardisation of a wider range of EU policy making and implementing processes. As a response to the corruption allegations, an attempt was made to depersonalise procedures within the EU and render them more technical in order to avoid misappropriation or the acceptance of personal advantage, leading to heavily bureaucratised standards. A Commission official in Brussels recounted how such standardisation is often cumbersome in the case of development cooperation with the partners:

> There is a lot of bureaucracy in European development cooperation, in particular when it comes to the disbursement of funds. [This] is a result of the éclat of the Santer Commission, then the EU has been trying to plug all possible loopholes by ridiculous and exaggerated means. It really reaches an extent where I am sometimes surprised that we can even sign a contract. The rules are so strict that some of EU member states wouldn’t even qualify. (Author interview, October 2008)

The quote is indicative of the wider perceptions of the EU’s role in Kenya. The EU’s way of conducting policy, both internally as well as externally, is determined by highly regulated spaces of interaction. Such regulation of interaction spaces is an integral part of the processes of regional integration in Europe and of defining the EU's external role (see Chapter 3). In different forms, these regulation processes are widely discussed, for instance in Moisio and Luukkonen's (2014) emphasis on routinisation in EU policy and decision-making processes, Falkner's (2013) pointing to the centrality of the rule of law for the EU's geopolitical credibility, or more general work on EU external diffusion of norms and policy models (see, for instance, Börzel and Risse 2012a). This trend towards regulation and routinisation has also long been an essential characteristic of African-EU relations. After half a century of trade relations and development cooperation, guided by long-term cooperation frameworks and strategy papers, the EU is primarily seen as a technical development implementer and trade partner. A political role and diplomatic relations were left to the member states. In a way, as one of Kuus' informants described, Commission staff have to remain technical – any attempt 'to say something [creative, different]' led to an attack by the member states (Kuus 2014, 129).

Kuus' work focussed on Brussels where the emphasis on process appears more pronounced than in the delegations. Informants also frequently explained how 'the closer you are to action on the ground, the less bureaucratic it all is. The closer you are to Brussels (but as a matter of fact also to your home capital) the more preoccupied with procedure you are' (Author interview, March 2014). Nevertheless, also in Nairobi the rough differentiation between collective, technical issues, on the one hand, and bilateral, political ones, on the other, was well noticeable. In particular during the first two years, Delegation officials did not necessarily see themselves in the driving seat for developing policies or political statements *vis-à-vis* the Kenyan partners, for as an EU diplomat clarified that 'it is not that the Delegation always comes up with the position and then tries to get that adopted by the member states'. Instead, the informant viewed it as the Delegation's key role to 'facilitate finding common ground' and to 'keep the communication channels open between us [the Delegation] and the member states but also between the member states themselves. It is a long process of gaining trust and confidence as an institution and when it comes to trust, personalities and the individuals are always very important!' (Author interview, November 2012).

The last aspect mentioned in the quote is central. Many informants voiced similar opinions about the role of individuals and personality in the conduct of diplomacy. This is a key point the EEAS is struggling with, and it is no coincidence that the EU diplomat articulating this point was part of the one third of EEAS staff that have come in from a national diplomatic service. It was the same diplomat that I quoted above who distinguished between the Commission's 'technical implementers' and the member states' diplomatic and political expertise. The Delegation's previous focus in Kenya on development and trade policy was different – one that could be approached through more technocratic and standardised procedures. The reorientation of EU external relations on foreign policy and diplomacy required a different, often personalised approach and sensitivity in interaction with individuals in key positions. The role of individuals and their personality, and more generally of human interaction, was frequently emphasised by the informants. A national diplomat, for instance, clearly articulated how

> [i]n light of the EU's ambition to regulate as much as possible and establish all kinds of structures, and a lot of people telling me that, in the end it boils down to who the person is you are interacting with, how do you judge the role of people, the human (individual) component in this process? Of course someone's personality is always important, in particular in foreign policy and diplomacy. These areas are just much more dependent on the people who fill them than, for instance, a protocol on how to establish the Common Market. (Author interview, November 2012)

Yet, after the corruption éclat of the Santer Commission in the 1990s, the widely adopted measures to avoid such scandals aimed exactly at banning such personalised policy conduct, resulted in an emphasis on policy procedures and further technocratisation and regulation of the spaces of interaction. EU policy was supposed to be de-humanised. In the world of diplomacy, however, the tact and instinct of people, and trust and confidence between individuals are essential. The EU's decades-long ambition towards institutionalising policy-conduct, its attempts to take faulty human nature (always prone to corruption and seeking the personal advantage) out of political interaction created a political entity with an unparalleled expertise in standardised, technocratic policy implementation but with an institutional culture unsuitable to a diplomatic world of subtleties, sensitivities and personal trust.

The spaces of such diplomatic interaction are inevitably different to those for the EU's previous role in Kenya in the fields of development and trade. As regards the *political* dimension, it is no longer possible to act merely as a technical implementer – even though this might be more in line with the EU's general preference for regulation, institutionalisation and standardisation of interaction spaces (see Chapter 3). In the diplomatic realm, however, social relations are of fundamental importance, i.e. precisely that human element that decades of forced bureaucratisation meant to take out of EU policy conduct.

Relations Between the Different Agents of European Diplomacy

> The EEAS could be conceptualized as an *interstitial organization* – that is, as an organization emerging in interstices between various organizational fields and recombining physical, informational, financial, legal and legitimacy resources stemming from organizations belonging to these different organizational fields. This interstitial status creates a situation in which there are different and sometimes conflicting organizational principles and practices introduced within the organization of the EEAS and different and sometimes conflicting sets of

> expectations in relation to the Service from actors within the organization as well as from outside. (Bátora 2013, 599)

Above, I have described the EEAS as a heterogeneous and composite diplomatic actor and EU collective diplomacy as a two-step diplomacy that is equally heterogeneous and composite. Key in enacting this diplomacy are the processes of negotiating relation between the different (political) entities whose interplay constitute the entirety of European External Action. This includes the relations between the different EU agencies in Brussels and between Brussels and third countries as well as the relation between the Delegation and the member states in third countries. Chapter 1 has already provided a brief introduction into how the EEAS has been established as a separate agency independent of the Commission and its established networks. In the EU's delegations, this separation between Commission and EEAS turned out to be complicating the practical functionality of those delegations considerably. An EU diplomat in Nairobi, for instance, explained how

> the institutional separation between EEAS and Commission was very bad, we waste so much time on the working arrangements of who does what. […] The EEAS was never intended as a separate agency. But it was turned into such through a decision by the Council. This was a wrong decision. We can't turn it back. Now we have to live with it and pay the price for the mistake. The way it works now is that it is two completely different agencies with two separate budgets. This is shit! (Author interview, March 2012)

Another EU diplomat emphasised the need for 'more coherence between the different silos [of EU external relations]. DG DevCo has the money but the EEAS wants to decide what to do with it'. The informant further lamented 'too many different tracks and too little coherences', however, also explained how the practicality of day-to-day interaction has significantly contributed to easing this separation: 'people work better together locally as they know each other (and go for lunch and coffee together). In Brussels there are a lot of turf wars and internal issues, people are still struggling over competencies, whilst in the Delegation here people are more pragmatic, they need things done together' (Author interview, January 2012). Other informants echoed this assessment arguing that the 'problem is pretty much only on a formal and administrative level. As regards substance and the cooperation in the delegation here, it is not a problem. This cooperation works well'. Nevertheless, it remains a 'cooperation'; EEAS and Commission staff 'cannot tap very much in each other's profile' – i.e. they cooperate well but with distinct fields of activity that are not supposed to be mixed. The only ones that are nominally entitled to cover both EEAS and Commission aspects are the Heads of Delegation as they 'wear double hats' (Author interview, November 2012). All other staff in the Delegations, including the Deputy Head of Delegation and the Head of Administration, are only mandated by either the EEAS or the

Commission, excluding them – at least nominally – from taking over tasks of the other agency.

With respect to the interaction between the Nairobi Delegation and member state embassies, the working atmosphere was described as constructive and positive by most informants. Yet, clearly, different individuals have different opinions on the general idea of European integration and of establishing a common diplomacy. One informant, a diplomat from a member state, was particularly critical of the entire process. Through our previous email conversations he was roughly aware of the topic of my research; when I eventually met with him for the interview, he was eager to express his opinion on the EEAS before I could even start asking questions:

> Let me say in general I am not happy with the coordination processes here. We are a small embassy covering a big agenda. We cover 13 countries, we would welcome more help from the Delegation for certain issues, in particular when dealing with the other countries where we are not based.
>
> […]
>
> It [the EEAS] is useless, because it does not do practical stuff and does not give us the assistance we could need.
>
> […]
>
> The more general question is if we need the EU at all? Look at the economic crisis. What happened to the ones in the Southern European countries who are responsible for it? What does the EU do about it? (Author interview, March 2012)

This informant was the only one of my interviewees who was questioning the EU in general, yet other national diplomats also voiced a critical stance on the state of cooperation between member states and the Delegation – or more generally on the idea of the EEAS. They were criticising a 'big waste of money' and 'a lot of duplication', as well as a 'lack of vision' (see above). The dominant attitude amongst member state diplomats towards the EEAS, however, was a largely positive one. Even though a diplomat from a smaller member state expressed a feeling that they 'have to work harder to be heard now; the rotating presidency had more of an open ear for the smaller ones' (Author interview, November 2012), she also emphasised that 'consensus-orientation is very strong; the smaller ones are not pushed around' (Author interview, February 2012). The 'strive for consensus-building' has been highlighted by all my informants; also those from smaller member states acknowledge that 'of course the bigger ones lead. But the smaller ones are listened to if they have something to contribute' (Author interview, March 2012). This articulates well with wider civilian and normative power debates on the EU and its general preference for multilateral and cooperative policy conduct (see Chapter 2). Instead of a feeling of 'being pushed around' by bigger member states, diplomats from smaller member states often expressed very positive attitudes towards the new structures of coordination:

> As a small member state we can always count on them [the Delegation] to cover things we cannot, we could also count on some of the bigger member states to do that. Whatever comes out of the EU Delegation we can usually live with. (Author interview, March 2012)

> We get a lot of information from the Delegation. In a way they really do the work for us. It is very helpful for the smaller countries as we rely on the minutes that are being sent around by them. As a very small country, I just cannot go and participate in every meeting for every sub-committee there is, but I am always informed as I get the minutes from the Delegation. (Author interview, February 2012)

> For small countries it is great, in particular for countries where we do not have embassies. I am very well linked to the delegations from the other counties we do not have our own embassy. I can't take care of all of this. Through the EEAS, I get the reports from these countries, this is really great! […]

> However, for the bigger countries I am not so sure. De facto they interact bilaterally with the partner countries, in particular the British Ambassador has much more influence here than the EU Ambassador. (Author interview, January 2012)

The last quote introduces a general trend of smaller member states being mostly content with the Delegation's assistance, in particular in terms of political reporting, and larger states using the collective vehicle when convenient, but displaying no hesitation in resorting to bilateral interaction.

EU Leadership on Foreign Policy

The lack of leadership on political issues on part of the Delegation is indeed a problem that many informants have pointed to, in particular in the phase between 2010 and 2012. This is tied to difficulties in adjusting to the new diplomatic role acquired through the Lisbon Treaty and the lack of experience as a foreign political agent beyond development and trade (see above). Informants from both small and large member states alike pointed to the Delegation's reluctance 'to play a more active role' (Author interview, March 2012). For diplomats from smaller member states the absence of a collective position and Delegation leadership often equalled the absence of diplomatic interaction – either because of the lack of particular interest or because of the lack of resources or means to advance such interests. For diplomats from larger member states, the absence of Delegation leadership only underlined the continued need for bilateral diplomacy. Emphasising the desire for a collective political role and the commitment to such a collective approach, a diplomat from a larger member state complained how

> [t]he Delegation shows no leadership on political issues. It follows an old style with no own opinion, they are technocrats, who coordinate things, order coffee, make sure the meetings run nicely. They facilitate things, but there is no leadership, no substance. In many ways it is the British who try to set the agenda, not the EU. The Delegation seems paralysed by the fear to do something wrong, it is just following. In light of this reluctance of political leadership by the EU, it might be better to do things bilaterally […]
>
> We should just collectively be much more vocal than we are (Author interview, February 2012)

Nevertheless, the changes that came along with the Lisbon Treaty with respect to the collective conduct of external relations, did influence the making of EU diplomacy in Kenya in many ways. Despite the frequent complaints about the lack of political orientation in the EEAS both in general (mostly directed at the HR/VP) and in Kenya, the 'coordination reflex' identified by Wessels (1982) more than three decades ago clearly received an impetus through the EEAS. It has moved beyond just 'coordination' to an active cooperation and institutionalised collective pursuit of foreign policy. It has also started to create collective spaces of *diplomatic* interaction, both physical and mental. A basic, yet significant, factor for the creation of such spaces of interaction was the physical change of meetings to always being hosted and chaired by the Delegation instead of by the rotating presidency. This provision of the Lisbon Treaty upgraded the centrality of the role of the Delegations and of Delegation staff. Both as regards the venue as well as agenda-setting and coordination, the Delegation became the central actor, instead of just 'one amongst many' (see also Bachmann 2015b).

During the course of the research project and the various field trips to Nairobi between 2010 and 2014, the evolution of the EEAS as the EU's key external relations agency was clearly observable – in particular on issues of foreign policy and diplomacy. In Kenya, the EEAS clearly pushed the integration of European external relations on the *political* level. Part of this evolution was a 'very smart strategy' of 'preemptive policy coordination' on the part of the EU Ambassador Lodewijk Briët and the growing section of political staff at the Delegation. Anticipating which topics and areas might become relevant for Kenya-EU relations, Briët discussed a list of such anticipated topics with member state ambassadors during regular meetings and pushed for the development of concerted positions on these topics. When any of these topics eventually did come up in relation to the Kenyan government, the Delegation could quickly release the previously discussed statement with the backing of the member states. Such 'preemption policy coordination' significantly reduced the Delegation's inertia and member states' reservation against common statements. Moreover, it also helped Briët 'to set the agenda and the tone; it prepared him to make statements on a wide range of issues and legitimated him to do so with the backing of the member states. It has armoured his position and set a framework for engagement with the Kenyan partners' (Author interview with member state diplomat, March 2014).

Another key factor for upgrading the EU's political role was the special situation of the cases of Kenyan President Uhuru Kenyatta and Vice-President William Ruto at the International Criminal Court (ICC) in The Hague – a situation that significantly facilitated the integration of diplomatic relations of European actors towards the Kenyan government (see Chapter 6). During an interview in November 2012, a member state diplomat, for instance, argued that 'at the moment you need to dig deep to find the drive for the EEAS. What you find is often motivated by shortsighted, very personal career interests. There is no collective vision' (Author interview, November 2012). In March 2014, the same informant explained how the 'Delegation's political role is [now] much stronger, it is also seen and recognised much more as the coordinator of the group' (Author interview, March 2014). Such an upgrade of the EU's collective *political* role was to be expected (or hoped for) after four years of trying to build a collective diplomacy, however, in the next chapter I argue that its key impetus was the special situation of the Kenyan ICC cases, not deliberate EU policies.

Chapter Conclusion: Heterogeneous Diplomacy and the Importance of Personality

> The EU possesses a foreign policy insofar its members agree upon it. (Carta 2012, 3)

> The focus on structure is inherent in the EU's administrative culture. (Author interview with member state diplomat, November 2012)

In this chapter, I have outlined some of the difficulties of *Europeanisation abroad*, that is the processes of building collective diplomacy between the different European partners in Nairobi (see Chapter 3). This first step of diplomacy was characterised by problems on part of the EEAS to move collective external relations beyond the EU's long-established role as a technical implementer of development and trade policy. The EEAS is struggling to acquire a political/diplomatic role as an institution that is, of course partially newly minted, but that is also emerging as the tentative endpoint of a half century evolution of the Commission's external action. Delegations around the world are not newly formed as EU Delegations; the Commission's delegations are transformed into EU Delegations under the roof of the EEAS. Despite its establishment as an independent agency, the EEAS therefore inherits, by default, some of the institutional culture and personnel of previous external missions of the European polity – with its own *Esprit de Corps* only emerging slowly (Benson-Rea and Shore 2012; Juncos and Pomorska 2013b). The 'inherent focus on structure in the EU's administrative culture' that the quote above points to has been expressed in more or less pronounced ways by informants from all backgrounds, including member states diplomats, EU officials and local cooperation partners. Asked about the relation between Commission and EEAS, an EU official in Brussels explained that 'as it always is when an organization is in the being, theoretically it is all defined, but practically there are difficulties. There are always situations where there are no standardized or regulated mechanisms *yet*' (emphasis added, Author interview, July 2012).

The 'yet' at the end of the quote, and the importance of structure and procedure attached to EU policy conduct described by most informants, are indicative of the wider strive towards the regulation of spaces of interaction alluded to in Chapter 3. This goes hand in hand with the, often lamented, underdeveloped political role and the lack of leadership on political/diplomatic issues in Kenya. Key to such roles are the personalities of the individuals filling the most important positions for diplomatic interaction – and after decades of de-personalising and institutionalising modes of policy conduct, this is certainly not the strength of the EU's *modi operandi*. The spaces of interaction that have long determined the EU's external action are poorly suited for personalised interaction required to play such a political role. They have traditionally replicated the EU's internal spaces of regulated and standardised interaction, with the level of bureaucratisation aggravated by the corruption scandal around the Santer Commission in the 1990s. The political dimension with its need for trust and tact does not sit easily with the EU's well-established expertise for technocratic governance. To fulfil expectations of becoming a more effective leading diplomatic actor, the EEAS will have to shorten its decision-making procedures, which requires the member states to subscribe to EEAS positions without being specifically consulted beforehand. More importantly, however, it will have to re-personalise policy conduct. It will have to move away from the image of being run by 'nameless, faceless bureaucrats' (Mathiason 2007, xii) by developing trust in its own personnel. Having interacted with several dozens of EU staff, it remains astonishing to me why European institutions have so little trust in their own 'highly accomplished and often charismatic individuals' as Merje Kuus (2012, 9) described the 'impressive group of professionals' she had worked with – a description I can only agree with.

Many informants explained to me how, in light of the lack of orientation on how to operate (in) the EEAS, the service functioned only because of good personalised relations with colleagues. When these relations functioned well, collective agency emerged; when they did not, there was no agency in the EEAS (RD Notes, January 2012). 'The character of how much we work together', a member state diplomat explained, 'very much depends on the people involved and to what extent they are cooperative'. Whilst this is a question of personality that can hardly be planned or standardised (in the context of the EU's ambition for regulated spaces of interaction), materiality also plays a role for the level of cooperation. The same member state diplomat continued to lay out that another important factor for the level of commitment to the EEAS was 'the size and the capacity of the embassies. Smaller embassies simply do not have the capacity to take part in *all* [meetings]. They are much more relying on collective approaches' (Author interview, November 2012).

Next to the personality of the individuals involved, such material factors of course retain their relevance and thus form an important part of critical geopolitical analysis (see Figure I.1 in Introduction). It is no coincidence that, in Nairobi, there is a general view that 'smaller member states are a lot more positive with the EEAS than the bigger ones' as an EU official in Brussels observed (Author interview, July 2012). A member state diplomat in Nairobi summarised the attitudes towards the EEAS, arguing that 'the EEAS is a great platform for smaller countries, but the bigger countries still do a lot of stuff on their own and sometimes even behind the back of the EU' (Author interview, February 2012). Due to different levels of capacity and resources (tied to national interests in different countries), the EEAS often is a necessity for smaller states; for bigger states it is just an additional tool in a larger toolbox of foreign policy instruments. Bigger member states use the collective approach when it is convenient; otherwise they act bilaterally. Similarly, bigger states are less dependent on the regulation of interaction spaces, as they have the capacity to act more autonomously. Yet the question of a country's size and power is always relative 'in a world dominated by continent-sized states' (Menon 2014, 20). To Menon, the 'fundamental problems' of building collective foreign policy 'lie not in the precise constellation of institutions in Brussels, but rather in the continued divergences between the interests of Member State anxious to retain their foreign policy prerogatives'. Member states, he argues, are faced with the choice 'between collective empowerment and autonomous decline. To date, they have largely plumped for the latter, emphasizing the pursuit of narrow national interests over effective multilateral action' (ibid.).

As for the making of EU diplomacy in Nairobi, the Lisbon Treaty's provision of centralising the functions of the rotating presidency at the Delegation was an important factor. Through this provision, the Delegation was ascribed a pivotal role. It provides the physical spaces of interaction in the form of the place of the Delegation premises as the venue for all meetings. In line with the approach on political spaces and the spatiality of the political outlined in Chapter 3, this place is politically empowering. It becomes a nodal position for the conduct of external relations, thus empowering the Delegation and its role in these networks of diplomatic conduct. Delegation staff function as hosts and organise interaction. They set the agenda, chair the meetings and coordinate the wider interactions of European diplomats in Nairobi. As such, through the physical relocation of spaces of interaction to the Delegation premises, the Delegation's role has shifted from being one actor amongst many towards acquiring the central and coordinating function.

The perceptions of this actor as central amongst its closest allies, the EU member states, varied. Most of the national diplomats were largely positive about the EEAS, one was rather sceptical and questioned the usefulness of the service but not of a collective approach as such; and only one of the informants had an outright negative opinion of the service to the extent of questioning the EU in general. All the other national diplomats were more positive than negative, critically pointing to issues needing improvement – mostly the lack of political orientation, bureaucracy and the inefficient use of money – but also emphasising the (potential) benefits of a concerted and collective diplomacy (see also Juncos and Pomorska 2013a). Without denying the problems of leadership in the EEAS, both in Brussels as well as regards political orientation in Nairobi, the member states' attitude towards it is interesting. On the one hand, the inertia and the lack of a clear direction and vision on political/diplomatic issues is frequently outlined as a key problem preventing the EEAS from becoming a more vocal collective diplomatic service that is more than the sum of its parts. As one national diplomat put it: 'we do not know what to expect from the Delegation' (Author interview, February 2012). On the other hand, it is precisely the reluctance of member states to trust the EEAS and subscribe to collective positions without being consulted beforehand that prolongs decision-making and creates wavering positions, since these are always negotiated compromises that can vary. A possible way out of this vicious circle is through the re-personalisation of diplomatic conduct – as has partially occurred in Nairobi through the ICC cases (see Chapter 6). For the EEAS this would entail focusing on individuals and content instead of structure and procedure as well as a clear political orientation on the part of the key individuals; for the member states it entails trust towards these key individuals and abandoning their insistence on prior consultation.

Chapter 6
Perceiving Collective Europe Abroad: Interaction with the 'Partners'

EU interaction with countries of the ACP-group is usually guided by regional and country strategy papers. The last Kenya-EU strategy paper in 2007 was to set the framework for cooperation for 2008–2013. Asked about a new strategy paper during an interview in Nairobi in March 2014, a diplomat involved in the negotiations responded with a smile on his face: 'We are working on it'. Further probed as to when it would come out, he clarified (still with a grin on his face): 'When it is finalized!' In June 2014, a new National Indicative Programme for the time period 2014–2020 was eventually signed between the Commission and the Kenyan government. The delay in deciding on a new strategy paper and the vacuum of directions for future cooperation is indicative of wider ruptures in Kenya-EU relations that have not only emerged but intensified with the new Kenyan administration under President Uhuru Kenyatta that came to power in March 2013. This rupture follows roughly a divide between Kenyan perceptions of European paternalisation and patronage on the one hand, and European perceptions of Kenyan corruption and poor governance on the other. I would argue that both are true.

Whilst the previous chapter has focused on the practices and processes of establishing the EEAS between different EU institutions, i.e. the first step of diplomacy and the mechanisms of *Europeanisation abroad*, this chapter is concerned with the second step of diplomacy and the mechanisms of *external Europeanisation* in Nairobi. It addresses the interaction with and the perceptions of the EU's external cooperation partners, such as the Kenyan government and other national and international organisations, by discussing how collective diplomacy functions towards the partners and to what extent the EEAS is achieving recognition as the key European foreign policy agency. The chapter extends the question of how the Delegation and the member states interact in a 'Europeanized' way to interaction with external partners; i.e. how is that new 'composite and heterogeneous' actor finding its niche in the existing realities of wider diplomatic conduct in Nairobi? And more so, how do external cooperation partners perceive of the EU as a collective diplomatic actor? It first describes the fields of interaction with the partners and the modes of conduct before it lays out the relation between multilateral and bilateral policy fields. It then briefly introduces the special situation in Kenya, created by the cases at the International Criminal Court (ICC) in The Hague against Kenyan President Uhuru Kenyatta and Vice-President William Ruto. The second part of the chapter deals with perceptions of the interaction and the EU's role in Kenya. It elaborates further on the differentiation opened up in the previous chapter of the EU as technical implementer or diplomatic actor. In this context it provides a detailed view of the pivotal role of the Head of Delegation in the process of building a political role for the EU. Finally it scrutinises wider perceptions of how the EU is seen in the context of other international actors in Nairobi's diplomatic circus and *vis-à-vis* the Kenyan government.

Operationalisation of the Second Step of Diplomacy: External Europeanisation

For the EEAS to develop a political role has not been an easy endeavour. It follows decades of interaction between Kenya and the Delegation of the European Commission in development and trade policy. It does not replace this interaction but instead supplements it through the addition of a political/diplomatic component. Political interaction, however, is of a different kind than development policy. During the interviews, in particular with Kenyan officials and NGO-leaders, I had to emphasise this distinction strongly and repeatedly. Many of the informants in this group conceived of cooperation with the EU exclusively in the domain of development or trade. Moreover, there is not a large number of individuals on the Kenyan side to approach for an interview that have experience in cooperating with the EU politically. In total I conducted 11 interviews with senior Kenyan government officials from different ministries, including the Foreign Ministry, the Ministry of Finance and the Office of the President. Whilst I did not succeed in approaching politicians on the ministerial level, the individuals interviewed held senior positions in different ministries, including the Head of External Resources, the Head of the ACP-EU division, the Director of the Europe and Commonwealth Division, current and former Kenyan ambassadors to the EU, etc. Moreover, I also organised a symposium on Kenya-EU relations, involving senior government officials, and was based at the Kenyan Foreign Ministry's *Foreign Service Institute* for a period of three weeks.

In explaining the background of the informants here, I want to underline that my research comprised senior governmental officials entrusted with Kenyan-EU relations on a *political* level; they were not lower rank implementers of specific development projects. Nevertheless, during all these interviews (as well as during the symposium and my affiliation with the foreign ministry), the 'development' context loomed large over the conversations and the informants' perceptions of the EU. It was little surprising that development-related topics were in the centre of interviews with officials from the Ministry of Finance, since this is the one coordinating development cooperation and, of course, budget support. However, even the six officials from the Ministry of Foreign Affairs that I interviewed were predominantly concerned with development relations; steering the interviews on matters of political relations was therefore a difficult task. This difficulty in gaining recognition as a *political* actor amongst its Kenyan partners also becomes evident in some of the interview quotes presented in this chapter. A diplomat from an EU member state explained how his country works 'with all: government, opposition, parliament, NGOs. We have our connections with political heavyweights from all parties. These are contacts that have grown over years. The EU simply does not have those contacts on a political level yet' (Author interview, January 2012). Other informants of both the EU and member states voiced similar takes on the situation, pointing to the 'Delegation's problems in accessing high-level contacts in the government' (Author interview, March 2014). Part of these problems is structural, i.e. related to the relative short time in which the EU has a political mandate; another part is political, i.e. related to controversy around the ICC cases.

Interacting with the 'Partner': Politics and/or Development?

In contrast to the interaction between the different European actors in Nairobi, described as *Europeanisation abroad* in Chapter 5, I focus here on *external Europeanisation,* i.e. the practices and processes of how Europeanised spaces of interaction are projected onto the relations with the EU's external partners. Cooperation extends to both the Kenyan government and to civil-society organisations (CSO). The dividing line between political and development cooperation is thereby a very fine one. As indicated above it posed considerable difficulties during the interview to maintain a focus on political issues. In a rather general opening question about the areas of political cooperation to an official from the Kenyan Foreign Ministry, the informant explained briefly that 'we cooperate in common values, such as human rights, the rule of law, democracy', before switching immediately to the more established field of development and starting a long monologue worth quoting at length here:

> Then of course there is also development cooperation. After the WorldBank, the EU is one of our biggest partners. But somehow it appears as if the EU is bringing more money to civil society. Maybe because they think CSOs are less corrupt. But you find the EU doing too much with NGOs and CSOs. The EU does not recognize that each country has a leadership that is elected and in the driver seat for development. This is where China beats the EU straight, because they realize that you have to work with the government. But still we work a lot with the European member states, with the EU directly, and also through the African Union and the African Peace Facility. The EU has helped a lot on Somalia, but they need to understand that we understand local things better than they do. They often forget this and need to listen to us.
>
> [...]
>
> But in general, there is too much on the political level. Now of course, the major preoccupation is with elections. You know, elections are nothing new to this country, we have held elections since independence. Only last time, there was a problem, but this was an exception. If you look at all the violence historically

> in state formation processes in Europe. Sometimes we tell your fellows you are not our supervisors, treat us as partners! Still our interaction with EU is intense. We share common values. Human Rights, rule of law, democracy. These are all values that are not exclusively European, but that we espouse equally. But there are also differences, for example when it comes to gay and lesbians. We are not ready for this. Europeans need to accept this. (Author interview, December 2012)

The quote illustrates many of the frustrations regarding the cooperation with the EU. The informant clearly sees EU cooperation with civil-society organisation as problematic and disrespectful of the government. By cooperating with NGOs, the informant implies, the EU undermines the 'elected leadership' and thus its own key principles of democracy. Instead, China is used as a contrasting example with more respect for Kenyan democracy due to its cooperation with the government. More generally, cooperation should be on development issues, the informant argues, because 'there is too much on the political level' including conditionalities attached to development issues. At the same time, he emphasises 'common values', such as democracy, human rights and the rule of law, highlighting Kenya's democratic tradition and denying the EU's claim to these values as 'European'. Yet by declaring such values as 'common', the informant accepts their universality and, implicitly, employs them as Kenya's moral benchmark against which the government's performance is to be measured. Such evaluation, however, is not to be done by Europeans. External involvement in 'internal' affairs is explicitly rejected and Europeans are perceived as 'supervisors', treating their Kenyan cooperation partners not as equal 'partners' but as inferior and in need of being told what to do.

Informants from civil-society organisations, however, often have a different opinion on this, even calling on the EU to hold the Kenyan government responsible for what it has committed itself to. During a symposium with Kenyan officials, NGOs, academics and EU and member state diplomats, a NGO representative highlighted the EU's role as a 'very significant player in solving the problems of this country [and] as a very visible mediator on the political level during the political crisis of 2007/2008'. To this informant it was through the EU's mediation efforts during the crisis and the support for the Panel of Eminent African Personalities headed by Kofi Annan, that the EU emerged as 'very, very significantly and constructively in the eyes of Kenya in a politically constructive manner'. Targeted at the Kenyan government, the informant argued that 'one of the problems in this country is that it quickly forgets its friends. […] And I would like the EU to continue reminding the Kenyan government that in our moment of need in 2007, there were very few people to stand with. But among us was the EU as one of the few friends'. More specifically, the informant continued to call on the EU on a political level to not let the Kenyan government 'off the hook' in solving the underlying problems in society that led to the outbreak of the crisis: 'I would expect that the EU would keep reminding Kenya that we have this little issue. The problems that lead to it are not exactly over, the commitment to solving these problems, which is a long-term commitment, has got to be carried through' (FG1, March 2012). The informant was the director of a NGO in strong support of the rule of law. The 'little issue' and the 'problems' he referred to relate to Kenya's long entrenched culture of political impunity as well as the underlying issues of kleptocracy and ethnic rivalries, often tied to unsettled land disputes (see Chapter 4) and seen as a key reason for triggering the violence.

For EU diplomacy, it creates the difficult situation of balancing its cooperation partners and achieving its political objective – insofar as one could be agreed upon (see Chapter 5 and 'first step of diplomacy'). Cooperation with government institutions clearly takes priority in Kenya-EU relations, in particular as regards political issues, but also with respect to development. Civil-society organisations are supported based on submitted proposals and a competitive system of grants awarded to successful proposals aligned with the EU's political objectives in the country. Contrary to the Kenyan government, CSOs actively call on the EU to increase its efforts on issues of 'good governance and the fight against corruption' (Author interview with CSO director, November 2012). At the same, however, the same informant also shared the government official's critique of Europeans lecturing Kenyans, arguing that 'we also want to see action, and not only lecturing. For example, they should not let corrupt people hide in Europe but actively work with us to get the corrupt elites and their companies'. He extended his critique on a European double-moral, frequently pointing to questions of governance in Kenya whilst simultaneously ignoring contentious economic issues perceived to be to the advantage of European companies:

> When it comes to human rights, there are often double standards. They [the "Europeans"] only talk about civil rights, but what about economic rights? For Kenya it is tiresome, they are always preaching about civil rights, but what about economic fairness and a level economic playing field? This is something you never hear about from the Europeans. Part of it is the accountability of foreign companies here in Kenya. Just as they call for the state to be accountable, companies also need to be accountable. For example when it comes to the exploration of oil, it needs to be transparent and accountable what Tullow [the British company holding the contract for exploration in Northern Kenya] does in this regard. If we, the CSOs push the state, they need to push their companies. (Author interview, November 2012)

The informant thereby hinted at a commonly perceived mismatch in the EU's external relations. Whilst the EU is widely perceived and accepted as an actor strong on human and environmental rights, its economic policy is equally widely perceived as oppressive towards and disadvantageous for African countries. I have elsewhere outlined this mismatch as a civilian/power dilemma showing such diverging perceptions of the EU as an often very positive geopolitical civilian model on the one hand and those of the EU as an oppressive geoeconomic power in the pursuit of self-interests on the other (Bachmann 2013a). This too indicates how external relations project in many ways the EU's internal spaces of interaction and the organisation of its political-economic system. Embedded in a highly normative political rhetoric, the European integration process has long pursued a broadly neo-liberal economic model often lacking sensitivities for unequal power structures or social concerns. In both internal and external policy orientation, the two facets cannot be disentangled; political normativity and economic (neo) liberalism have long gone hand in hand as part of the European integration project. Whilst the latter might initially have been intended as a means to tie former political foes together through the creation of economic interdependencies, the temptation of capitalist enrichment has obfuscated discernment of inequality and often normative concerns. As regards external perceptions, the EU's appeal for its broadly civilian political orientation, therefore, is generally accompanied by (neo)imperialist perceptions in the economic realm and as regards interference in domestic affairs.

The Special Situation of Kenya's ICC Cases

The perception of such illegitimate interference is often related to questions of good governance, corruption and the rule of law – aspects that have long been central to European external relations. Whilst the European emphasis on these aspects is generally welcome amongst civil society organisations, it is a frequent issue of controversy in the cooperation with the Kenyan government. The particular situation of the cases of President Uhuru Kenyatta and Vice-President William Ruto at the International Criminal Court (ICC) was thereby a main topic of contention between 2012-2014. Even though there was a deliberate attempt on the part of the EU not to position itself in favour of any of the candidates prior to the presidential election in March 2013, support for the ICC-processes was outspoken. At the same time, the EU has been equally outspoken against a closure of the cases or their return to a Kenyan or East African court, as favoured by the indicted Kenyatta and Ruto. It was clear that, if elected, cooperation with the suspects would be seen as difficult. A member state diplomat clearly articulated how 'we all [EU member states] have similar takes on these issues. We all disrespect most of the Kenyan politicians. And we all want to believe that the next election will be better. We all write the reports and sell it to our governments. But that does not mean we believe in our reports' (Author interview, November 2012). Similarly, another member state diplomat recounted how in the second half of 2012, the meetings of the EU political counsellors were determined by the upcoming elections and the ICC cases, arguing that 'it will be very problematic if Uhuru [Kenyatta] or Ruto are elected. If this is the case we do not know how to treat them' (Author interview, November 2012).

When Kenyatta and Ruto were eventually elected a vocal 'EU-bashing' in the words of a member state diplomat (Author interview, March 2014) emerged, framed around the EU's alleged support for the other candidate, Raila Odinga, and portraying the ICC cases as a Western trial against Africa(ns). After the elections, the European ambassadors requested a meeting as a group with the new government, which was refused by Kenyatta. Instead he conducted a number of bilateral meetings; first with the ambassadors of the US and the UK, then also with the EU Ambassador and some national ambassadors. There is a general realisation amongst European diplomats that this was a 'punishment' for particularly outspoken ambassadors in the context of the ICC processes (Author interviews, March 2014). Also Kenyatta's refusal to swiftly accredit the newly appointed ambassadors of France, Germany and Italy as well as the hesitant invitation to the celebrations for Kenya's fiftieth independence anniversary in December 2013 for some of the largest development partners, such as Denmark, Finland, Sweden, and the Netherlands, has been seen by diplomats 'as clearly related to the ICC cases' (Author interview, March 2014). Explaining the effects of this new situation for collective diplomacy, a member state diplomat articulated how

> within the new government, there is a decidedly anti-western tone and resentment against the West – of course related to the ICC cases. So, sometimes joint action is even counterproductive because it triggers immediately the reflex of the West imposing things on Kenya. We still do it sometimes, mostly for problematic issues where we know it will be a confrontation. In a way, we as the member states are then hiding behind the collective approach because no single country is being exposed as the main critic. (Author interview, March 2014)

Also other member state diplomats have pointed critically to the quick release of common public statements, not as critiques of the collective approach, but in the context of advocating a more 'silent diplomacy' in order not to fuel those voices that are only waiting to interpret any 'Western' statement as a moralising inference in domestic affairs, for instance in relation to homosexuality (Author interviews, March 2014).

However, it is also in the context of Kenyatta's 'diplomatic snub' (Omondi 2013) towards some of the EU member states, that the Delegation, and in particular the EU Ambassador Lodewijk Briët, moved more to the centre of European diplomacy. Several member states adopted an 'essential contact only diplomacy' towards the Kenyan government, waiting for a resolution of the ICC cases (Author interviews, March 2014). Also in light of the British 'doing their own thing anyway' (Author interview with member state diplomat, March 2014), the vacuum caused by the temporary absence of three of the five big EU countries' ambassadors, was to some extent filled by the EU Head of Delegation. Briët, whose diplomatic role 'has significantly strengthened' as a result of these events, emerged as a 'key spokesperson and the key person to address on behalf of all of the EU' (Author interview, March 2014) as a member state diplomat explained who was previously one of the most critical of the collective diplomatic approach and the Delegation's role in Nairobi. Other member state diplomats echoed this view, arguing that 'in a way, the entire issue around the elections and the ICC has united us more as a group. We met very frequently, like once a week or so, sometimes even more often' (Author interview, March 2014). On the one hand, the special situation of the Kenyan ICC cases has clearly contributed to politicising the EU's role and its perception in Kenya. On the other hand, they continue to pose a significant problem to the relations with the Kenyan government and have further entrenched the perception of European moralisation against Africa(ns). The entire election campaign of the Kenyatta/Ruto presidency had been framed around a rhetoric of Western influence in internal African affairs and of positioning the ICC as a Western institution for putting Africans on trial. Their success is proof to the persistence of such sentiments in Kenyan society. This, however, is not an issue exclusively associated with the EU's collective political role in Kenya, but is more generally targeted against the West.

Heterogeneous Agency, Heterogenous Perceptions

Just as the EEAS is a composite and heterogeneous diplomatic actor, so the perceptions of this actor are equally heterogeneous. In Chapter 5, I alluded to the different perceptions of European diplomats on the EEAS as a political/diplomatic actor or as a technical implementer in the field of development cooperation. This rough differentiation is retained in perceptions of the EU amongst its cooperation partners in Kenya. In addition, the role of the EU as an economic actor needs to be taken into account. At the centre of my work, however, are questions that enquire about the EU's role as a political actor and the processes of acquiring diplomatic standing in Nairobi's wider diplomatic arena. Even though I was not primarily interested in development cooperation, this kept recurring as a key topic in conversations and interviews, most likely for two interrelated reasons. First, the EU has conducted development programmes in Kenya for several decades. Second, development aspects and politics are not clearly separable when a large share of Kenya-EU interaction revolves around development issues. However, in certain aspects of EU interaction with the Kenyan government, the separation between development and politics is artificially maintained. The following excerpt from my research diary reproduces a conversation between two Delegation officials in Nairobi that I was part of and indicates how the development/political divide is far from being clear.

> Helena and Michael were talking about the processes of disbursing budget support to Kenya and that still there are about €8m not disbursed yet. The initial figure of €29m was reduced to €21m due to corruption charges. Helena was not happy with the process because of the behaviour of DG DevCO in Brussels. It was a political decision to reduce the money and to not disburse it yet – also based on pressure of the member states (in particular Germany, the Netherlands and the UK). The EU is the only donor in Kenya to do budget support, however

> in Brussels the concerns of the member states as regards corruption are shared. The problem is that DevCO tries to cover up this political decision with technical matters, i.e. that there are technical issues that delay the disbursement of funds. DevCO apparently does not have the guts to say that it is for political reasons and because of concerns about corruption. The Delegation here has given the OK for the practical disbursement of the funds; the technical issues DevCO raises do not exist! Helena said that she views the Delegation's role in terms of checking these technical issues but that it is Brussels who needs to make the political decision about what to do.

It might be important here to note that Helena was a Delegation official from the European Commission. She clearly saw her role as a technical implementer (as Commission staff) and not as somebody to get involved in a political matter such as reducing cooperation due to political differences, i.e. corruption charges. Moreover, the excerpt also illustrates the Commission's reluctance to articulate its political opinion and adopt a political role, despite the inherently political process of injecting funds into the budget of a different state. The Commission (through DG DevCO) thus continues to treat political processes as technical matters. It is trying to take the *political* away from politics – a process that James Ferguson described as an 'Anti-Politics Machine' more than two decades ago (Ferguson 1990). An official from the Kenyan Ministry of Finance clearly observed the political nature of budget support, arguing how 'budget support has been very unpredictable recently. This is a big problem. There are a lot of conditions attached. One of which is governance and governance is always a matter of perception' (Author interview, March 2012).

Budget support has long been favoured by the European Commission as a tool of development cooperation, yet the Commission rejects the political role that comes with it. Nominally, the separation between the institutions is clear with the Commission as the institution conducting EU development and trade policy in Kenya and the EEAS in charge of political and diplomatic relations. This separation, however, is impossible to maintain in practice and also not seen by the Kenyan cooperation partners. A Kenyan official clearly articulated how they 'do not care about the internal divisions and the internal set-up. The EU is the EU, we don't care about a division between EEAS and Commission or whatever else. To us, EEAS or Commission is mainly a matter of labelling within the EU. It has no effect on how we deal with them' (Author interview, March 2012). Asked about how the introduction of the EEAS had changed their work, another Kenyan official responded: 'Not at all, it has not changed at all, we don't even notice it. It just does not matter. The EEAS is an internal EU thing, for us it has not changed a lot. Our relations and cooperation is written out in the Cotonou agreement. That is what counts for us, not Lisbon or the EEAS' (Author interview, July 2012). As a consequence of this 'invisibility' of the EEAS, its political role is hardly noticed. In the eyes of Kenyan officials, the introduction of the EEAS (after all the most substantial restructuring of EU external relations in its history) has not resulted in any change in cooperating with the EU or an upgrade of its political role. Instead, there was a 'rather pronounced political role of the EC in 2007/2008 when we had some political gymnastics in this country'. The informant, a senior Kenyan diplomat, was referring to the post-election violence, which occurred before the establishment of the EEAS. He continued:

> but other than that the EC's role has been very specific in assisting for economic development and doing some reforms (economic and in the judiciary). For the multilaterals, getting involved in political issues is out of tune. It is like with the Worldbank. Politics is not part of the multitlaterals. In general I think the multilaterals (EC, Worldbank, etc.) should stick to development and the political aspects should be left to the ambassadors of the US, UK, Germany and so. The multilaterals should not duplicate this. Politics is not their business. (Author interview, March 2012)

Even though I frequently asked about the EU's and the Delegation's role in Kenya, the informant kept on referring to the EC, meaning the European Commission. In no way do I intend to judge this, given the complex heterogeneity of different EU agencies, however, the dominance of the Commission's development cooperation programmes in Kenyan officials' perceptions clearly underlines how subdued the EU's political role still is in relation to its overshadowing development programmes. EU officials are well aware of this, confirming how 'in Kenya, the EU is first and foremost seen as a development partner. It does not play the big political role. Real and important "political" deals are still done mostly on a bilateral level' (Author interview, November 2012). In the pre-Lisbon era, this was partially due to the ambiguous 'relationship between political mandate and technical expertise in the production of EU policy. The former is to come from the member states while the latter is to be supplied by the union's civil service' (Kuus 2014, 74–5). Policy implementation then followed a pattern whereby 'once a basic political compromise is struck among the member states, the issue is shifted to the realm of technical expertise to be provided by the commission (ibid., 118). These observations accurately described the relation between intergovernmental alignment and supranational implementation before the Lisbon Treaty. They also reflect well this relationship for the first few years of EEAS' existence. From 2013 onwards, however, this distinction became much less clear in Kenya.

Personalising the EU's Political Role: The Head of Delegation

The special situation of the 2013 Kenyan elections and the ICC cases of President Kenyatta and Vice-President Ruto have changed the situation significantly whereby the EU's political (lead) role has become much more pronounced. 'The big question', according to a member state diplomat, is 'if the EEAS should be more like UNDP or more like the US State Department. It has one advantage over the state department and that is the level of control of the Head of Delegation in the missions abroad and the sheer amount of money they are responsible for'. The diplomat continued to argue that, due to their double hats (Commission and EEAS), EU Heads of Delegation have more competencies than national Heads of Mission. It is in their power to 'delay the signature of development projects. They cannot just abandon them, but they can delay them. This is more power than national ambassadors have, also due to the sheer volume of the funds' (Author interview, March 2014). The conversation with the diplomat took place within the wider context of the EEAS' ambition to upgrade the EU's political role. He further argued that 'the driving 10 per cent of the EEAS seem to have realised its potential to become an independent actor' and that the new EEAS staff that has arrived at the Nairobi Delegation 'clearly interprets its role much more political' than the Commission staff at the Delegation who 'continue to act as technical implementers'. However, probed further if Delegation staff are being permitted to play a stronger political role on the part of the member states and on the part of the Kenyan government, the diplomat affirmed that 'on the part of the member states, yes. The Kenyan government does not really understand the internal divisions within the Delegation' (Author interview with member state diplomat, March 2014).

All diplomats interviewed in 2014 confirmed that the Delegation's political role has been significantly upgraded over the course of 2013 – occupying much more of a coordinating and leading position in European diplomacy than before. The key factor thereby was the role of Lodewijk Briët, the EU Head of Delegation in Nairobi, who, according to a member state diplomat 'is acting much more political than before. He is much more present in the media, his presence has tripled' (Author interview, March 2014). The reasons for this are to be found in the special situation of the ICC cases. In light of the 'essential contact only diplomacy' of some member states and the simultaneous absence of the French, German and Italian ambassadors in 2013, Briët played a key role in filling the vacuum of high-level European diplomats in Nairobi. He thereby also became 'a prime target' for the new Kenyan government in its anti-Western rhetoric 'as he was often the one making statements on behalf of the EU member states' (Author interview, March 2014). Whilst this put the Delegation's direct interaction with the Kenyan government under strain, his political profile and visibility rose substantially at the time. A member state diplomat observed: 'if the government does not talk to you, you have to talk louder and through the media so that they read about it. And, in addition, sometimes the member states hide behind the EU Delegation when it comes to conflictual topics they do not want to address themselves' (Author interview, March 2014).

Giving member states the opportunity to 'hide' behind the collective approach, is a role Delegation staff diplomats appeared to appreciate since it was elevating their political profile. Such 'hiding', an EU diplomat posited, is 'what is called Realpolitik. You use a political tool when it is convenient to you. The member states use us when it is convenient to them in order to advance their interests' (Author interview, November 2012). The key person to 'hide' behind on political issues was Briët as 'the one spearheading the European position. He clearly is the accepted speaker for us [EU actors] on common positions. He is not just one amongst many other ambassadors' as a member state diplomat explained (Author interview, March 2014). It was a combination of these special circumstances, created by the ICC-cases that contributed to an upgrade of the Delegation's political role in Kenya. The EU thereby emerged as 'definitely more than just development aid; it does play a political role. It is also realized in the media and in the Kenyan public as fulfilling a political mandate – often in a negative way, though, for pushing Kenya to the ICC' (Author interview, March 2014).

Yet despite this often negative publicity, the political dimension became an increasingly visible and pronounced component of EU external relations in Kenya as part of this special situation around the ICC cases. Member state diplomats repeatedly attributed this to Briët, who was frequently described by informants as 'doing a very good job in presenting the full spectrum of the EU in Kenya, political and economic, not only development. Considering he also comes from a development background, he has clearly taken on the political and diplomatic role, providing leadership in this area' (Author interview with member state diplomat, March 2014). Other member state diplomats echoed this view of the EU's elevated political visibility and role largely being attributed to the person of the EU Ambassador, describing how he now has a strong 'political voice, not just that of a technical implementer of development policy' and how amongst the senior European voices in the media 'you see him [Briët] and Turner [the British High Commissioner], but Lodewijk is much more visible than Turner'. In general, according to a member state diplomat, the EU Ambassador might be a

> boring leader, but a very visible one and one that is highly respected and listened to amongst both the member states and the Kenyans. He is much more recognized as *the* collective European voice than before. One year ago, if he would say something, some national ambassadors would also say something. It would be the same thing, but they would say something just to be recognized, too. Now Lodewijk's role is much stronger. He is very much accepted as *the one* speaking on behalf of the EU. Also, the countries that were previously very outspoken about the ICC are more muted now. He is much more accepted – also by the big member states. But I am also sure that if he goes too far, the member states would intervene, both here and in Brussels. (Author interview, March 2014)

The last sentence in the quote resonates with Carta's (2012, 3) assessment that 'the EU possesses a foreign policy insofar its members agree upon it'. However, the specific circumstances of the elections and the ICC cases in Kenya, have created a situation whereby the EU Delegation, and in particular its Head, Ambassador Briët, succeeded in positioning itself as a 'strong and respected' political voice that member states can, mostly, not only agree with, but also hide behind as a conveyer of confrontational messages and statements.

Development After All: Positioning the EU Between Patronage and the Need to Disburse Funds

Formally, Kenya-EU relations remain guided by long-term cooperation frameworks, such as the Joint Africa-EU Strategy, the Cotonou Agreement as well as regional and country strategy papers (see Chapter 1). This focus on long-term and institutionalised cooperation is part of the EU's wider geopolitical ambition to regulate spaces of interaction with its partners. The Commission's general preference for budget support as a tool of development cooperation, framed in long-term strategic frameworks, is also indicative of this trend. Along with such orientation, however, comes a reduction in flexibility and the possibility to react to unforeseen changes, or the immediacy of a political situation, as was seen with the ICC cases. During a previous research project in Nairobi, a British development worker pointed to the Commission's preference for 'long-term commitment and cooperation frameworks determined for five year periods. But it is also less flexible than bilateral' (Author interview, February 2008). Cooperation, in particular development funding, is fixed in five-year strategy papers and once the budgeting cycles are approved by the European Parliament, the funds have to be dispersed within the envisioned time frame. A member state diplomat explained that EU institutions are always 'under the pressure to spend money. […] It is not well regarded in Brussels if a delegation does not spend all its allocated money. The African cooperation partners know this; they know the money is going to come anyway, regardless of what happens, of what they do, how they act and how they treat the EU partners' (Author interview, January 2012).

This reflects the EU's preference for institutionalising interaction. The EU acts as a large bureaucracy; in over 50 years it has developed a profound expertise in shaping institutions, regulatory mechanisms and legal frameworks that govern interaction in an institutionalised, bureaucratic, *regulated* way. It is incapable of interacting in an environment of (military) power politics and rapid changes of political situations. Such fixed frameworks and inflexibility deprives EU development cooperation of an important tool of power projection. In contrast, a member state diplomat argued, 'China is very different here, because if they are not happy with something, they just stop paying. It is a dictatorship and can act a lot more flexibly in this regard'. Such leverage to withhold funds, largely absent to the EU's development portfolio, results on EU cooperation as being taken for granted and 'much more publicity' being given to China as a partner. The member state diplomat continued that

African governments 'need to schmooze the Chinese to get their money whilst they know that the European money they get anyway. We just can't threaten to withhold funds as the Chinese do'. (Author interview, January 2012).

This applies not only to development cooperation. Also in light of the 'diplomatic snub' surrounding President Kenyatta's refusal to meet European ambassadors after the election, a member state diplomat argued that 'the Kenyans know that there will be no consequences, the money will come anyway. Kenyans learnt that it is almost always business as usual' (Author interview, March 2014). Similar attitudes have frequently been expressed by European diplomats, lamenting the lack of attention on the part of the Kenyan government and the prominence enjoyed by Chinese diplomats and projects in the media and amongst the government. For most Kenyan informants, however, it was only a logical consequence to engage more strongly with China. There is an acute awareness of European countries 'also being on the decline' and possibly not being able 'to give us the kind of assistance they used to do before, given that they are also just emerging from their own debt crisis. So it leaves us with no option but also to look East, where it is easier to get access and all those things without much conditions' (FG1, March 2012). Similarly, the director of a key civil society organisation, who was generally very critical of the Kenyan government and corruption in the country, outlined clearly how

> [former president] Kibaki has made a bold move to engage with China. But now people see roads and say this is what we wanted all along, they do not want to be lectured on corruption and human rights all the time. Of course there is corruption, but also visible development. People do not care about the corruption because they can still see something happening. And with the West, there were 30 years of ODA [official development assistance] and you don't see any results. What happened to all this money? So if there is corruption with China, but you still see results, you can't say anything.
>
> […]
>
> Kibaki has made the decision to invest in infrastructure. He has been looking for partners in this area. China has been a key partner for infrastructure. So he made the decision to engage with those who are willing to cooperate with him on this issue he has made a priority. If you are not willing to do that, do not complain if you do not get attention. (Author interview, November 2012)

Whilst the informant welcomed Chinese cooperation in terms of infrastructure, he was equally critical of an omnipresent Chinese engagement in Kenya, pointing to the 'Chinese invasion, and this is clearly what it is, but people don't see it yet' and arguing that 'it will take a lot of time until people here will hate China, it will come for sure, but not yet'. To the EU, he recommended taking 'a more balanced view' and 'not spend so much time telling Kenya not to work with China. They should not demonise China, but work with them' (Author interview, November 2012). The Chinese hands-off policy of non-inference and without conditionalities thereby stands in immediate contrast to persisting perceptions of European paternalisation towards African countries (Men 2011; see also Power, Mohan, and Tan-Mullins 2012; Bachmann 2013b). Informants from NGOs or civil society organisations, such as the one above, often took a balanced view of both Chinese and European involvement in Kenya, arguing that 'in general we want to spread external involvement and not be overly dependent on one single source. And thereby gain some space for ourselves' (Author interview, November 2012). Problems in the Kenyan government and with Chinese involvement were clearly seen, including the EU's role in pointing to issues of corruption, democracy and human rights and in holding the government accountable to promises made in these issues. However, simultaneously perceptions of European arrogance and patronage continue to loom large. A Kenyan public intellectual, politically close to the government, explained how

> [t]he EU is very irresponsible as it comes with preconceived notions of Kenya and Africa and the way it should be and what it should do. EU conceives a Kenya that it wants and defines it accordingly. It then uses these measures as the parameter to evaluate Kenya on. Benchmarks for Kenya are predefined by the EU. This is a very problematic issue as regards dignity and respect. Dignity in dealing with somebody else is not a question of how much money is in the pocket. We know of all the problems we have, and in our political system, ourselves. But we want partnership, not patronage.
>
> […]
>
> There is a complete disconnect between the EU (and some of its member states) with the power elites in this country and how they work. The EU has its own vision of Kenya, which is very different from what is actually going on here. Kenyan elites have other visions of this country, these visions are also very different and very heavily debated. There is a big discourse going on in the country, which the EU often doesn't understand because it does not understand tribal structures underlying politics in this country. (Author interview, November 2012)

The aspect the informant was referring to are fundamental aspects about *external Europeanisation*, i.e. practices and processes through which the EU seeks to project its internally developed 'European' spaces of interaction to the relations with others. He criticised notions of the EU predefining what its relations with Kenya should look like, but also how Kenya as a country was supposed to function, including its political system. The lack of an understanding of 'African realities and conditions' voiced by the informant was shared by many other Kenyan informants, from both the government and civil society organisations. He also pointed to the underlying tribal structure that determined politics in this country, that had always been present and were possibly aggravated by a perceived domination of the Kikuyu community, and that were the driving forces for the post-election violence in 2008 (see Chapter 4). Any account of Kenyan politics that ignores these dynamics is incomplete. According to the informant, Europeans have no comprehensive understanding thereof, yet they seek to impose a political-economic model unsuitable to a Kenya determined fundamentally by tribal logics that Europeans do not understand.

The assessment of European imposition of their positions, however, was seen differently by many of the informants. Another NGO director, more critical of the government, explained how in negotiations with Delegation personnel, there is always a wish to 'strive to find a consensus. They do not impose their view on us. The process is usually driven by our own priorities, which is good. They try to inform themselves of others' opinions as a basis for forming their own opinion' (Author interview, February 2012). Beyond the larger political criticism of corruption, government officials also pointed to the Delegation's consensus orientation in project-based negotiations, arguing that 'negotiations are not a one-way street, we talk and work together' (Author interview, March 2012). 'If there is no consensus', an official at the East African Community involved in the EPA negotiations explained, 'our negotiations will go on – as is currently the case with the EPAs' (Author interview, November 2012).

Perceptions, of course, are in the eye of the beholder and his/her experiences of past cooperation with the EU. Naturally they vary widely, however a general trend can be observed that, also amongst Kenyan cooperation partners who are critical of the government, outspoken European patronage and moral 'preaching' is widely rejected. It is, as one informant put it, a question of dignity and respect that many Kenyan informants see as missing in the political tone – leading to the widespread perceptions of a continued hierarchical donor-recipient relationship. This does not apply to perceptions of the practical interaction with Delegation officials. On this level of direct interactions, all Kenyan informants, from both the government and civil-society organisations, confirmed a cooperative and consensus-seeking character of interaction.

More generally, in the wider context of the 'essentials only diplomacy' related to the ICC cases, consensual interaction on concrete development or trade related projects and cooperation programmes seems to have moved more into focus. 'Cooperation with Kenya' a member state diplomat remarked, has 'shifted maybe a bit away from principles and bigotry, more towards practical issues' (Author interview, March 2014). Another member state diplomat explained how 'at the moment the key forum for exchange with the government is the development group'. The diplomat referred to a high-level meeting between Vice-President Ruto and the European Heads of Mission in preparation of a more concrete development forum, involving all donors. According to the informant the meeting was 'very good and constructive, it also included issues outside development, such as human rights and also some political issues. Communication with the Vice-President and the atmosphere were 'indeed very constructive and good' (Author interview, March 2014). Also another member state diplomat described

> the new government, both in the President's and the Vice-President's camp [as] very pragmatic; they are interested in business. So our communication at the moment is pretty much issue based. In a way we are dancing around the ICC cases. There is a way around this and we can still interact with the Kenyan government, but confrontation is never far – we have also heart things from

> government officials like 'we know who our real friends are' referring to India and China. (Author interview, March 2014)

In light of such fragility of communication routes and the political dissonance from the ICC cases, the EU's emphasis on consensual interaction, mutual 'ownership' and the identification of the Kenyan side with the particular project or agreement is both pragmatic and essential. On the one hand, in a situation of 'essentials only diplomacy', it constitutes a way of achieving constructive interaction with the Kenyan government, unspoilt by creating additional perceptions of unilateral imposition. On the other hand, it is essential in light of the EU's insistence on long-term frameworks and agreements for cooperation. Any unilateral agreement, perceived as disadvantageous by the Kenyan partner would be fragile and under a constant threat of cancellation or re-negotiation. Such uncertainty does not fit in with the EU's objective of creating predictable, institutionalised, *regulated* spaces of interaction. At the same time, however, the using of development issues for reinvigorating dialogue reproduces the image of the EU as a technical implementer of development and does little to bolster a vaguely emerging diplomatic role.

Chapter Conclusion: Emerging Diplomacy Beyond Regulation

A regulation of the spaces of interaction of EU diplomacy in Kenya was far from being realised in the political climate overshadowed by the ICC cases against President Kenyatta and Vice-President Ruto. A result of this special situation, however, has been the significant upgrading of the EU's political role and visibility in Nairobi's diplomatic arena – mostly associated with the person of the EU Head of Delegation, Ambassador Briët. Through the 'very close coordination' of EU member states around the time of the elections in March 2013, Briët emerged as the key conveyer of European diplomacy. Even though Briët, and by extension the European Delegation in Nairobi, became a key target of the new Kenyan government's 'EU-bashing' and 'anti-Western resentment' (which complicated diplomatic interaction), the situation also positioned him, and by extension the EU, at the centre of focus of public perception in the political/diplomatic realm. This was clearly a change to the rather subdued political role of the Delegation before the elections. In 2012, an EU official pointed to the 'limited political expectations' held towards the Nairobi Delegation and argued that it 'would not be a bad political role to play to function as door openers for the member states and to facilitate contact and exchange' (Author interview, November 2012). In 2014, the Delegation's political role had significantly changed. Member state diplomats clearly positioned the 'EEAS more as an overarching institution. They make first drafts, circulate those, we discuss and modify them, but then the Delegation is also the one conveying and articulating them' (Author interview, March 2014). This upgrade of the EU's political role in Kenya was not a result of the grand scheme of European External Action planning. It was also not a result of successful *external*

Europeanisation, i.e. the external projection of internal (regulated) spaces of interaction. In fact spaces of diplomatic interaction are far from being regulated in this context, as they barely exist in light of the ICC controversy. The upgrade of the EU's political role resulted from the special situation created by the ICC cases. It resulted from personalised conduct of diplomacy on the part of the EU Ambassador, based on the member states' (willing and unwilling) diplomatic reticence and their simultaneous entrusting Briët as the 'key person to address on behalf of all the EU'.

During the phase of 'essentials-only diplomacy', several member state diplomats have pointed to the importance of the development group or other project-specific encounters as fora for interaction with the Kenyan government. This interaction has been described as 'very good and constructive'; both sides consider it to be a welcome way to deal with and get to know the 'partner', while 'dancing around' the ICC controversy. The Commission's traditional policy fields of development and economic cooperation thus remain guided by both sides' preference for long-term agreements and stable cooperation frameworks. On the part of the Kenyan side, this provides a guarantee that cooperation will also continue in times of political flux and differences with the EU – a situation that several European diplomats criticised as a lack of a political tool to withhold funds. On the part of the European side, the preference for long-term agreements is part of its broader geopolitical objective of institutionalised and predictable conduct of external relations within clearly defined, *regulated*, spaces of interaction (see Chapter 3).

Perceptions of the EU amongst the Kenyan partners thereby vary between government officials and civil-society organisations. Whilst the former prefer to see the EU exclusively as a development and trade partner and are highly critical of the EU's cooperation with civil-society organisations, the latter view the EU as an important political partner in 'reminding' and pushing the government to fulfil its commitments with respect to human rights and good governance. Both, however, share a strong rejection of European patronage and moralising and criticise European 'double-morals', in particular when it comes to questions of migration, framing conditions for a 'fair' economic relationship, and Chinese involvement in Africa. They also share a more positive impression of the EU, that of the Delegation's consensus-orientation in project-specific interaction, articulating clearly how 'they do not impose their positions on us'.

With respect to the EU's collective political role, the special situation of the Kenyan ICC cases was an important factor assisting the emergence of the Delegation as a 'loud and respected' political voice in Kenya, backed by the diplomatic community of its member states. This politicising of the EU's role, a key objective of the EEAS, was not achieved through policy decisions in Brussels or the deliberate attempt at the external projection of internal spaces of interaction. The formal provisions entitled upon the Delegation and the Head of the Delegation through the Lisbon treaty provided the structures and the mandate for collective diplomacy, but those structures remained hollow until the special situation in Kenya filled them with life. It was the member states' interest in closer cooperation and collective conduct, induced by the ICC cases, and the role of the EU Ambassador that enabled the emergence of collective diplomacy. Such unplanned and ad-hoc situations are clearly not part of the expertise of the EU's political portfolio, but depend on competent and entrusted personnel. Planned processes of engineering *external Europeanisation* through EU agencies thus remain limited to policy fields of traditional Commission competence, i.e. development and trade. There, Europeanised regulated spaces of interaction determine Kenya-EU relations – largely in the interest of both Kenyan and European partners. In these policy fields the EU continues to act, and be seen, as a technical implementer of development aid.

Whilst this articulates with the EU's key geopolitical objective of creating regulated spaces of interaction (see Chapter 3), it sits uneasily with its ambition to acquire political clout. Political recognition in Kenya and the Delegation's role in the centre of collective European diplomacy were achieved decidedly differently. The attempts of the early phase of the EEAS (2010–2012) of *external Europeanisation*, of the deliberate projection of regulated interaction spaces onto the conduct of diplomacy, have failed. Standardisation, bureaucratisation, and 'depersonalisation' are unsuitable for diplomacy. Diplomatic conduct requires trust, confidence and personal relations of those in charge of enacting political relations. Such factors cannot be institutionalised and regulated. This creates the irony that the EU's strive for regulated spaces of interaction as the preferred (only?) framework within which it can effectively exercise power (see Chapter 3), is unsuitable for the political role it aspires. For acquiring political clout, a re-personalisation and flexibilisation of EU External Action is required.

The critical geopolitical approach employed here has been instrumental for making sense of both Kenyan perceptions of the EU and the mismatch between the EU's preferred geopolitical space of regulated interaction and its aspired political role. All of the four couples of factors (historical and spatial, political and economic, local and international, representational and material) are at play here. The historical background of colonialism, but also Kenya's post-independence evolution and its ongoing ethnic divisions, immediately shape contemporary politics in Kenya (see Chapter 4). In addition, several components condition Kenya-EU relations. The ICC cases, the Kenyan election campaign, the volume of development aid, the long disputed negotiations around the Economic Partnership Agreements (EPAs), the European migration regimes, the Chinese role in Kenya, and the EU's insistence on particular regulated spaces of interaction are key factors that shape Kenyan interaction with and perceptions of the EU.

These contextualities allow, at least partially, for developing an understanding of ongoing perceptions of European patronage amongst Kenyan informants – both governmental and non-governmental. Also the eventual emergence of a collective European diplomacy in Kenya is framed in these contexts. Yet a decisive factor was the perception of the EU Ambassador as a trustworthy person amongst member state diplomats. For researchers, any approach seeking to study Kenya-EU interaction and Kenyan perceptions of the EU that lacks sensitivity to these contextualities is incomplete. Remote and purely textual analysis cannot grasp such factors. I am far from claiming global comprehension here, yet my ethnographic approach of being based in both European and Kenyan institutions gave me access to relevant individuals. By using ethnographic methods, I have aimed to capture important subtleties and insights on both sides in order to develop a better understanding of the mundane practices of *making* EU diplomacy.

Chapter 7
What Should the EU Do? Expectations of the EU as a Geopolitical Actor

The EU's role as a geopolitical actor is often assessed in terms of civilian and normative power approaches (see Chapter 2). These debates have generally had both a descriptive and prescriptive character by trying to analyse the EU's role as well as laying out visions of what the EU should be or do on the global stage (Nicolaïdis and Howse 2002). The roles ascribed to the EU have been fairly consistent over the decades of European integration, from the civilian power notion of the early 1970s to the more contemporary notion of normative power Europe (NPE). Of course, these often idealistic notions have also been challenged by those denying the EU the status of any kind of power due to its inherent (political and military) limitations (Bull 1982; Maull 2005; Hyde-Price 2006; Menon 2014). However, an aspect that almost all of the debates on Europe's collective geopolitical role share, is the positioning of the EU as a geopolitical actor that is 'better' than others (Diez 2005; Rogers 2009): better than the USA and Russia during the Cold War, better than the USA during George W. Bush neoconservative revisionism, better than Chinese autocracy, etc. The reason for this alleged superiority has always been found in Europe's supposed more civilian, normative, moral, 'fair' orientation in comparison to other global powers.

More concretely, in a comprehensive account on the EU's global role, Bretherton and Vogler (2006, 56–7) have outlined three complementary roles for the EU in the world: the EU as a model for regional integration, the EU as a promoter of normative and civilian values, and the EU as a geopolitical counterweight to the USA. As others, also Bretherton and Vogler are both descriptive and prescriptive, tracing patterns in the EU's global role since the construction of a collective external role in the early integration steps as well as idealising specific roles for the EU to play. The roles identified by Bretherton and Vogler are also largely reflected in the results of a previous research project I had conducted between 2006–2009 (see Chapter 2). The project focussed on European development policy in Kenya, but also contained a section inquiring about expectations regarding the EU's wider geopolitical role. Most prominent in the responses in this earlier project were expectations that the EU should function as an alternative geopolitical model of civilian power and regional integration and promote this internationally (the first two roles of Bretherton and Vogler). Also Bretherton and Vogler's third role, the EU as a counterweight to the US, was mentioned frequently. Other respondents expected the EU to manage globalisation and world trade and to act as a partner for African countries (see Figure 2.1).

I refer here to this earlier project and Chapter 2 in order to illustrate how expectations of the EU's role as a global and development actor have changed between the two research projects, mainly due to the outbreak of the financial crisis in Europe. The expectations presented in this chapter are based on research conducted between 2010 and 2014. Whilst the two previous chapters focussed on the processes and practices of establishing collective European diplomacy, this chapter introduces expectations held towards the EEAS and the EU. It includes the expectations of all the interview partners as well as those obtained through a symposium and the focus group discussions. It also teases out different expectations depending on the background and institutional affiliation of the informant or source of information. What is illustrated is therefore a presentation of very subjective opinions of the informants on what role Europe *should* be playing in their view. However, through this subjectivity it illustrates how expectations vary depending on affiliation, background and viewpoint. In so doing, this chapter covers three types of expectations: a) European diplomats' and Kenyan cooperation partners' expectations of the EEAS as the new structure of a collective European diplomacy; b) expectations of the EU's role in Kenya and the wider East African region; and c) expectations of the EU's wider geopolitical role in the world.

What to Expect from the New Institution of the EEAS

> We are fully in support of integration in areas that are EU competences. In others, we have our national interests and the relation has to remain one amongst good friends.
>
> [...]
>
> I think it should stick to the issues that are Union competence and not engage in other issues, such as security issues. (Author interview with member state diplomats, March 2014)

> I think everyone needs to know their place. Ours is not in defence and security. We need to be aware of that. (Author interview with EU diplomat, March 2012)

Individuals from different backgrounds and with different institutional affiliations naturally have different expectations regarding the new institution of the EEAS. The interviews with European diplomats and Kenyan officials, however, interestingly revealed how such different expectations do not necessarily follow a clear pattern of institutional affiliation. Rather, they were heavily influenced by each individual's personal political orientation and, occasionally also, their respective government's Europhile or Eurosceptic orientation in the case of member state diplomats. While perceptions of the EEAS' usefulness to member states followed a rough differentiation between big member states making use of the EEAS when convenient and small member states relying on its support more strongly, in terms of expectations what the EEAS *should* do no such patterns appeared.

A general expectation that most European diplomats shared was that the EEAS should raise visibility for European engagement in Kenya – both for collective engagement and for individual member states. In light of the EU's role as the largest trade and development partner to Kenya, the lack of visibility is also tied to a fair amount of jealousy in comparison to the prominence of China in public media and the (perceived) ease of Chinese access to high-level government officials. A key problem thereby is the lack of any possibility of using the threat of withholding funds as a political tool to exert pressure on the Kenyan government, 'resulting from an inherent inertia and a lot of bureaucratic hurdles' in European development policy (see Chapter 6). According to a member state diplomat, a prime example of this is the European funding for the African Union stabilizing mission in Somalia: 'the Somali government completely depends on AMISOM, which is completely paid by the EU. If AMISOM wasn't there, the Somali transitional national government would break down in a few weeks. Still they fool us wherever they can. If we would threaten to withdraw funding for AMISOM, we could exercise much more influence there' (Author interview, January 2012). Several member state diplomats expressed their expectation on the EEAS to raise the EU's profile and make its engagement in Kenya (and the region) more visible. Vague statements that the EEAS 'should advance our European interests' (Author interview, January 2012) or that it should 'give Europe a bigger voice, a collective voice' (Author interview, January 2012) are commonly shared. Some diplomats were also adding more concrete expectations to the strengthening of a collective voice, such as 'support democracy, human rights, good governance, etc.'. (Author interview, November 2012) or the call to tell 'its own story; a short and simply story of how to build peace. It should look at the history of European integration and spread this message. It is a success story for building peace' (Author interview, February 2012). Many of these expectations resonate with the dominant geopolitical discourses for the EU's role in the world (see Chapter 2). Civilian and normative notions appear throughout. However, also calls for exercising more power through strengthening collective action articulate not only with idealistic civilian and normative power debates, but also respond to realist critiques thereof.

In terms of more concrete functions to fulfil, opinions diverge more widely. Particularly diverse were responses to the question of whether the EEAS should act as an overarching institution bundling the member states diplomat services under a common roof, or if it should act as a supplementary agency assisting national services in fulfilling their diplomatic mandate. Perceptions of this question also change decidedly with the controversies around the Kenyan elections in March 2013 (see Chapter 6). Before the elections only the most Europhile diplomats saw an overarching function for the EEAS. Even EU diplomats regarded the EEAS as 'one tool in the box of the foreign policy instruments of the member states. I think it would be a very nice accomplishment if the EEAS can enrich the arsenal of member states' foreign policy tools'. Asked for her view of the expectations of member states for the EEAS, the same EU diplomat responded that, in her opinion, member states view the EEAS 'as a secretariat that can take over certain duties for them. […] But I do not think they expect a lot from us at the moment' (Author interview, November 2012). Before the elections, this attitude of the EEAS as a service to assist those of the member states was held prominently amongst many national and EU diplomats alike. A member state diplomat, for instance, summarised a common take amongst the European diplomats:

> I think the EEAS should look for places where it can be useful in addition to what is being done by the national services already. There is a lot more potential for the EEAS to act both collectively, but also more independently as a service that identifies areas where it can be of added value and can make an impact. They need to believe that they can do something, identify the areas and then do it. At the moment, it is a bit process heavy and it takes too long until any decisions are made. It can choose the areas where the EU has leverage to make an impact and not so much focus on developing simple statements. Its leverage can be in quite a few areas. Trade for example, since the Commission negotiates access to the entire EU market. This is a huge leverage. Maybe eventually also immigration and visa issues. It needs to identify areas where the EU is better than its member states and I think these are quite a few areas.
>
> […]
>
> I think the EU is often seen as more neutral here in Kenya. Also when it comes to elections, the EU is seen more neutral than US and UK. Those come with a big influence, but also with a heavy baggage. (Author interview, November 2012)

The member state diplomat hinted at the key role the Delegation, in particular EU Ambassador Briët, would be playing in the context of the elections that were only a few months ahead of the time the interview was conducted. However, like the EU diplomat before, the member state diplomat also positioned the EEAS as a service to assist and support member states, while calling on the EEAS to identify niche areas of operation in which it could exercise its advantage over national services. The thought of an overarching service is not present in either of these statements. Moreover, a decided political role for the EEAS is also not envisioned by the informant. Collective action should be limited to areas where 'the EU has leverage'. Those areas are predominantly those of Commission competence, for instance development and trade. An 'intrusion' of the EEAS into the diplomatic domain is not foreseen by the informant.

Opinions on this, however, diverged. Another member state diplomat suggested for the EEAS to move towards being more than just one among the many European national diplomatic services, arguing that '[the EEAS] should take over certain functions currently performed by the member states. At the moment it is still more coordination, but in future it will be more steering and taking over certain functions'. The informant continued that in many fields, 'de facto the cooperation is already very close; also on consular affairs', but also emphasised that, in his view, it was highly unlikely that 'there will be collective consular sections any time soon. The issue of immigration is too sensitive and member states do not want to entrust others in this regard' (Author interview, November 2012). Roughly one and a half years later I interviewed the same informant again. In the context of our discussion about the effects of the Kenyan elections and the ICC cases, he then clearly described how the 'taking over of certain functions' had progressed in particular in terms of political communication with the Kenyan government and the conveying of collective demarches. Asked about the overarching or supplementary character of the EEAS, he clearly positioned the service as 'an overarching institution' 'spearheading' European diplomacy in Kenya (Author interview, March 2014). Chapter 6 lays out how, by March 2014, he was not alone in this opinion. Other member state diplomats similarly described the Delegation's political role in Kenya as an 'umbrella role', 'providing leadership', or as the 'coordinator of the group' (Author interviews, March 2014). Even though different opinions about the EEAS' overarching or supplementary character persist (and will probably continue to do so), there was general positive consensus amongst member state diplomats towards the EU's elevated political role – either as an overarching 'umbrella' or as a stronger supplementary voice in support of national services. This backing for the EEAS on behalf of the member states is essential for its functionality and the way it is perceived by its cooperation partners. On a more general discussion of the EEAS' role, and in particular Catherine Ashton's in it, an official at the East African Community (EAC) argued that

> it is always easier to have one interlocutor than several, if it is mandated and relevant! Otherwise, it is not worth having more European institutions to deal with. What is there is enough. If European countries want to be more integrated, they need to give her [Ashton] more clout. Otherwise it is irrelevant and not interesting for us to deal with. (Author interview, November 2012)

Amongst the EEAS' Kenyan cooperation partners, there was a similar attitude. The change in the EU's external relations machinery did not impact on their interaction with the Delegation, as they regarded it as an 'internal EU thing'. Consequently they did not voice particular expectations towards the new institution. On the contrary, there was a plethora of different expectations amongst all of the informants interviewed as part of this research regarding the EU's role in Kenya and the East African region as well as the EU's wider geopolitical role as a global actor. Those will be introduced in the following two sections.

The expectations are also summarised in Figures 7.1 and 7.2 (below), which illustrate a quantified grouping of the most common responses and an explanation of the categories whereby the lighter, upper part of the figures represent the numbers of responses given by the EU's cooperation partners, i.e. mostly Kenyan government officials and NGO representatives, and the darker, lower part the number or responses given by European diplomats (from both the EU or member states). Whilst the number of European diplomats was higher in the interviews (36 compared to 20 cooperation partners), the number of (Kenyan) cooperation partners was significantly higher in the focus groups and the preceding short questionnaire (71 to 4). The graphs presented here (and in Chapter 8) combine responses from those data sources and thus show a higher number of responses from cooperation partners than from European diplomats. In total, 100 informants responded to the question on expectations of the EU's role in Kenya and the wider East African region (22 European diplomats, 78 (Kenyan) cooperation partners) and 69 informants to the question of expectations of the EU's geopolitical role as a global actor (20 European diplomats and 49 (Kenyan) cooperation partners)

Expectations of the EU's role in Kenya

The EU has long held deep and broad relations with Kenya. Kenya is a key 'partner' country for the EU, not least due to its lead role in East Africa, and the Nairobi Delegation is one of the most important outposts of the EU (see Figure I.1 and Chapter 4). For Kenya, the EU continues to be the biggest donor of ODA (official development assistance) and the largest trade partner. It is therefore no surprise that also expectations on the EU's role in Kenya focused on development (37 responses) and the generation of economic growth, mostly through trade (34). Other important issues mentioned frequently are related to good governance, peace and security (32) as well as education and capacity building (24). Moreover, 16 responses also wanted to see a less neo-colonial role of the EU in the region, mainly referring to perceived interference in domestic affairs (see also Figure 8.3). The EU's role as a promoter of regional integration (22) still appeared as an important factor amongst the responses given, however this did not compare to the centrality of 'regional integration' in an earlier survey prior to the outbreak of the financial crisis in Europe (see Figure 2.1).

The largest category, *Development*, underlines the role of the EU as a development actor in Kenya and the East African Region. For both Kenyan and European informants, development assistance constitutes an important part of Kenya-EU relations and is expected to continue. This is hardly surprising given that development has been a cornerstone of relations since the establishment of the European Delegation in Nairobi in 1976. It also squares with the argument made in Chapters 5 and 6 of the EU being seen primarily as a development actor or technical implementer of development projects – in particular before the Kenyan elections of 2013 and the controversy over the ICC cases which led to an upgrade in perceptions of the EU's political role. A Delegation official observed how 'in Kenya, the EU is first and foremost seen as a development partner' (Author interview, November 2012). A large part of the EU's development role in Kenya is expected to be on infrastructure projects, in particular road construction. This too results from a long engagement of Commission support to infrastructure projects in Kenya through the European Development Fund and the strategy papers. This expectation is shared amongst Kenyan and European informants. Many European diplomats referred to infrastructure development as a key pillar of collective development policy, based on the long experience of Kenyan-EU cooperation in this area and also the relative ease of its implementation as a largely apolitical, more technical process with high visibility and presentable impact: 'The EU should do big infrastructure projects here to help Kenya improve economically' (Author interview with member state diplomat, March 2012). Similarly, Kenyan officials from the Ministry of Finance repeatedly argued that the 'EU should stick to doing infrastructure. Budget support is very tricky, we even reduced it because it was so problematic. We recommend the budget support share in the EU assistance to Kenya to be reduced' (Author interviews, March 2012).

I briefly alluded to the problems with budget support in Chapter 6. Tied to certain conditionalities around good governance, inserting funds into another state's budget is a highly political process. The conditionalities attached to budget support (or other forms of development assistance) are of course an ideal means for the projection of its preferred spaces of interaction through *external Europeanisation.* By attaching conditions to the provision of aid, those conditions can be formulated in support of the EU's geopolitical objectives, i.e. the regulation of interaction spaces. Despite the complexity involved, the Commission continues to treat budget support as a technical matter of development cooperation, causing frustration amongst its own staff responsible for its implementation, but also amongst the Kenyan partners due to the conditionalities and the unreliability in cases of non-compliance. Infrastructure projects, on the other hand, are indeed more technical and apolitical and can often be conducted without the direct involvement of the government (provided that the relevant titles and licenses have been issued). Moreover, they are much more visible to both the local population as well as to a European electorate increasingly demanding information about what happens to taxpayer's money in third countries. A constructed road is easier to 'sell' to a critical European audience, also with respect to the argument of making markets easier accessible for European consumers (flowers, tea, coffee) and European companies than largely invisible budget support. Informants from different backgrounds called for the EU to 'lessen its involvement in the international politics of countries but increase its engagement in infrastructure' or to 'do something about the improvement of infrastructure. Good infrastructure will spur economic development' (FG questionnaire, March 2012). Conditionalities were thereby often seen as problematic by Kenyan informants because they were largely perceived as a paternalising interference in domestic affairs. However, at the same time many Kenyan non-governmental informants expected the EU to 'encourage governance, transparency, accountability' as 'very key in areas of development and improving participation' (FG questionnaire, March 2012).

Another large category describes Kenyan expectations for the EU in the areas of *Trade and Economy*. For Kenyan informants, this was even the largest category, more important than *Development*. For European informants, despite acknowledging that the 'support of development, business, enterprise and investment in Kenya is also in our own economic interest' (Author interview, November 2012), trade and economic cooperation did not appear to play a key role. Despite its importance in the region, in terms of the overall external trade of the EU, Kenya remains negligible. On the contrary for Kenya, the EU is the largest trade partner. Consequently, for Kenyan informants trade with the EU is a key expectation, as they called on the EU to 'open more opportunities like trade for the East Africa region' and to 'open up more to Kenyan commodities as well as those from the East African region' (FG questionnaire, March 2012). Even the negotiations for the Economic Partnership Agreement (EPA) were not necessarily seen in a negative light any more, contrary to earlier perceptions of the EPA (Bachmann 2013b). Trade is seen as an area of cooperation on a more 'equal' level than a hierarchical donor-recipient relationship extending benefits from one side to the other. It is therefore also seen as an area of potential mutual benefit. Indicative of such sentiments amongst Kenyan informants, a symposium participant argued how 'we are looking forward to favourable outcomes of the EPA negotiations. They will trigger more investment, infrastructure, and capacity building. We need a change of perceptions; moving away from a hierarchical relationship to one of partnership. We need more reciprocity' (FG1, March 2012). During the phase of economic liberalisation under the Kibaki administration (2003–2013), Kenya has moved from being classified as a Least Developed Country (LDC) towards being classified a low-income country. As such it cannot be part of the Everything-But-Arms (EBA) initiative that guarantees duty and quota free access to European markets for products from LDCs. The expiry of these preferential non-reciprocal trade agreements under the Cotonou agreement thereby has to be seen as a sign of Kenyan economic emancipation and a chance to build a more 'equal', 'reciprocal' base for Kenya-EU relations around trade and economic issues – beyond a hierarchy-reproducing monster called *Development* (Kiamba and Bachmann 2015).

The third of the larger categories, *Good Governance, Peace and Security*, is a rather heterogeneous category, grouping expectations for the EU to function as a guardian of good governance, human rights and democratic standards, but also pointing to the EU's engagement in peace building missions, in particular in the Horn of Africa (AMISOM). For both, the EU is seen as a credible actor given its broad civilian power orientation and commitment to governance standards and human rights as well as its history of transforming a continent of warfare to a regionally integrated civilian union. This articulates with the EU's main geopolitical discourses of civilian and normative power (see Chapter 2). Respondents also often combined the two aspects arguing that the EU 'shall support Kenya and East Africa in their quest to maintain security and also join hands in fighting for people's rights' or expecting the EU to 'cease support for corrupt, tribalistic, nepotistic regimes. That respect for human rights be a key component before aid is awarded' (FG questionnaire, March 2012). The EU is seen as embodying a certain value orientation around good governance and human rights issues, however it also carries a double standard through its support of questionable regimes. With regard to the specific situation of the Kenyan ICC cases, civil society organisations often called on the EU to be a 'strong pro-ICC' voice and to 'remind Kenya of the need to resolve the underlying problems' that led to the outbreak of the post-election violence (FG1, March 2012). In general, the EU's role as a security actor appeared much more prominently than could be expected; its role in the peace building process in Somalia was well recognised in Kenya, especially given Kenya's own military engagement in Somalia. Several Kenyan informants emphasised the importance of the EU's engagement in this context, requesting that the EU 'fund more projects that are geared towards peace building and peace education as a process of mitigation and prevention of violent conflicts in the East Africa region' (FG questionnaire, March 2012). This slightly ambiguous category displayed a rough distinction between European and Kenyan informants, with the former mostly emphasising the EU's role in the area of good governance and human rights, and the latter calling for the EU to support peace building efforts and act as a security provider.

European support in the areas of *Education and capacity building* was expected much more on the part of the Kenyan informants than by European informants. The category is related to expectations in the area of *Development*, but framed less hierarchically. Informants argued that the EU 'should seek to foster mutual benefits by encouraging trade, (academic) exchange programmes and keep off politics' and that it should do 'capacity building for Kenyan public and private entities to manage their own affairs while benefitting from the resultant foreign exchange' (FG questionnaire, March 2012). Overall, responses in this category were less concerned with the extension of development aid and rather focused on the support for self-empowering exercises through scholarships, exchange programmes, as well as knowledge and technology transfer.

The category of *Regional integration* featured prominently especially amongst European informants. Perceptions in this regard have changed substantially since the outbreak of the financial crisis in Europe. The entire regional integration narrative, previously a key source of Europe's appeal to the rest of the world (Fioramonti and Poletti 2008; Lucarelli and Fioramonti 2010; Bachmann 2013a), has come under severe strain as the crisis revealed some of the problems that come along with a large, often inert polity full of heterogeneous interests – often in the context of solving the financial problems in Southern Europe. Nevertheless, the model of a peaceful regionally integrated union continues to emit a certain appeal, in particular amongst European diplomats, who argue that the EU should tell 'the story of building peace' amongst former enemies (RD notes, March 2012). This, too, replicates the general tone of the civilian and normative power discourses and the creation of regulated interaction spaces (see Chapter 2 and 3). Kenyan informants recognised the peace building effect but were generally more concerned with issues of generating economic growth and development through regional integration, expecting the EU to 'help the East African countries in formulating policy that can be able to build and sustain the East African community' (FG questionnaire, March 2012).

The smallest category, *less neocolonial*, is nevertheless a highly important one. It refers to perceptions of the EU as a neocolonial, neoimperial power that is paternalising Kenya through aid conditionalities and moral 'preaching' about good governance and human rights, thereby illegitimately interfering in Kenya's domestic affairs. In a way, this category stands in contrast to expectations of the EU in the area of *Good Governance, Peace and Security*. Often informants also take both stances, calling on the EU to be a guardian of good governance and human rights but also rejecting any indication of interference. In a rather drastic way, an informant demanded that the EU 'stop meddling in Kenya and the East-Africa Region like, for example, with the ongoing ICC cases. It should stop neocolonialism or else the citizens will rise against the EU, especially Britain' (FG questionnaire, March 2012). Somewhat more diplomatically, a Kenyan official explained that

> [w]e are partners, at least that is the word that is often used in the Cotonou agreement and also in other places. If we are equal partners or not, that is a different thing. I think we can all expect mutual respect for each other and the partnership. Not the paternalistic way of talking down. We need dialogue and mutual respect. That is done in direct interaction, not in the public. (Author interview, July 2012)

The lack of respect and of equality in Kenya-EU relations has been frequently mentioned, however the consensual interaction on a practical direct level has also been acknowledged by other Kenyan informants. There seems to be a general difference between neo-imperial perceptions on a political macro-level, fuelled by the current controversies over the ICC cases, and good and respectful working relations on a direct, practical level of interaction (see also Chapter 6).

Expectations of the EU's Global Role

Chapter 2 outlines some of the key discourses on Europe's global role, whereby in particular the three roles identified by Bretherton and Vogler (2006) of the EU as a model for regional integration, as a promoter of normative and civilian values, and as a geopolitical counterweight to the USA, are relevant. Whilst a previous research project (2006–2009) identified expectations that can roughly be structured along Bretherton and Vogler's three roles, those roles appear less dominant in this later research, in particular amongst perceptions of the EU's cooperation partners. Most of the categories on the EU's role in Kenya (Figure 7.1) also appeared in the expectations of the EU's global role (Figure 7.2). There are, however, significant differences with respect to the ordering of the categories. Moreover, the category of *less neocolonial* did not emerge independently out of

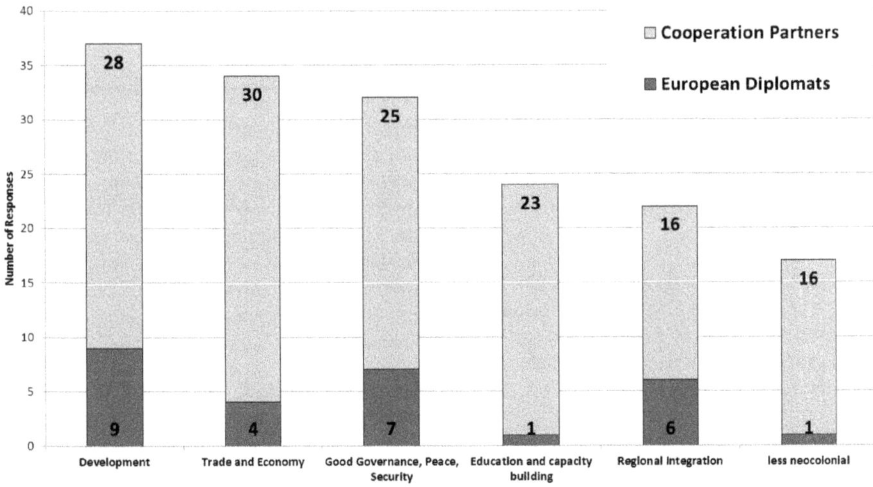

Figure 7.1 Expectations on the EU's role in Kenya and East Africa

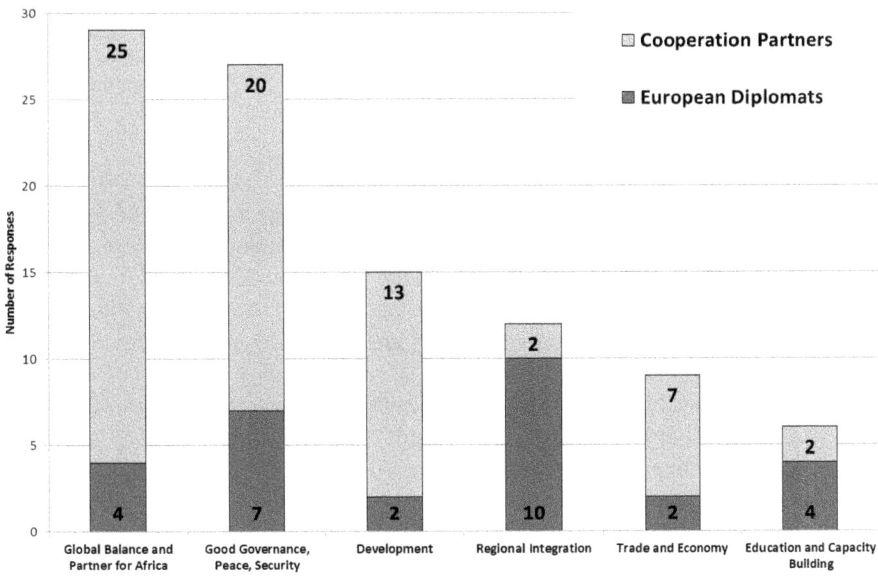

Figure 7.2 Expectations of the EU's global role

Textbox 7.1 Explanations of the categories of Figures 7.1 and 7.2

Trade and Economy
This category refers to issues of trade, economic relations, the management of the global political economy and facilitated access for East African countries to European and global markets.

Development
Responses in this category referred to a broad range of issues in the context of development cooperation, such as financial and technical assistance, infrastructure, reduction of poverty, health policy, etc.

Good Governance, Peace, Security
This category includes both 'soft' and 'hard' political aspects. The former refers to issues of good governance, democracy, accountability and transparency, human rights, etc. As regards the latter, respondents mentioned a stronger European role in the areas of peace and security, conflict prevention and resolution, including military components.

Less Neocolonial
This category summarises perceptions of the EU acting in a neo-colonial manner, dictating the terms of Kenya-EU relations and a strong inference in domestic politics.

Education and Capacity Building
Responses in the category were often also linked to the category of development; however, they were focused more directly on the promotion of education in the Global South, mostly through financing education, research and training locally, but also through exchange programmes, as well as technology transfer and assistance in capacity building in a variety of sectors.

Regional Integration
Respondents advocated for the EU to support regional integration initiatives, mainly with respect to the East African Community (EAC) but also as regards the African Union (AU).

Global Balance and Partner for Africa
These responses were strongly normative and called upon the EU to use its (economic) strength to create a more 'just' global order, in particular with regards the functioning of the world economy, and to act as a partner for African countries – in a non-paternalising way. One respondent nicely summarised the given responses: 'Empower Africa but not to re-colonise Africa'.

the responses; instead a new category grouping the highest number of responses emerged around expectations on the EU as an actor standing for *Global Balance and Partner for Africa* (29 responses). The other categories are smaller, but refer to issues similar to those described in the previous section (see Figure 7.1): *Development* (15 responses), *Regional Integration* (12), *Trade and Economy* (9) and *Education and Capacity Building* (6). Expectations of the EU's role as a promoter of regional integration are significantly less pronounced than before the outbreak of the crisis (see Chapter 2). Also, of the informants expecting the EU to promote regional integration internationally, 10 are European diplomats and only 2 are cooperation partners of the EU.

Responses to the largest category, *Global Balance and Partner for Africa,* largely reproduce expectations that are often outlined as idealised roles in official strategy papers of African-European partnerships on a global level, such as for instance in the Joint Africa-EU Strategy (JAES). They invoke a partnership between the EU (as a collective polity but also with respect to the member states) and African countries and regional organisations. They also refer to common goals such as the creation of a 'fair' global economic playing field or for the EU to function as a partner for African countries' integration into the global economy. As early as 1993 Christopher Hill outlined four international roles for the, then, European Community (EC). Among these were 'managing world trade' and the attestation that the EC 'remains the principal interlocutor with the poor majority in the UN' and should therefore function as the 'principal voice of the developed world in relations with the South' (Hill 1993, 310–11). And indeed, Kenyan informants frequently voiced their expectation that the 'EU assists Kenya under the ACP-framework to integrate into the world economy' (Author interview, March 2012). This was often tied to the EU's role as a geoeconomic power with a considerable influence on the global economic architecture. Kenyan and European informants alike argued that what the 'EU should do globally is world financial regulation' (Author interview with member state diplomat, February 2012) and that the EU is 'a very powerful economic bloc and has a role to play in stabilising the world economy' (Author interview with Kenyan official, March 2012). Also focus groups discussants often expected the EU to 'stand for economic fairness and good practice. China should not be allowed to dump sub-standard products in Africa. EU should also stand for democracy and good governance in the world' (FG questionnaire, March 2012). The last expectation also hints at the EU's standing as a (alleged) guardian of 'universal' values and a voice for poorer countries as indicated by Hill (1993).

The third role outlined by Bretherton and Vogler (2006), that of the EU as a counterweight to the US, appears in this context, though it is modified. It is not so much the US that is the focus; instead it is China that is increasingly perceived as a superpower that needs to be balanced out. The EU is not necessarily seen as on par with these superpowers, but at least economically it is seen as a global power with considerable influence. A member state diplomat, for instance, argued that the EU

> should be a mitigating force in future brokering between big forces. A little bit like a big Switzerland. The EU's idea for peace and peaceful cooperation is great. It should be the world's dominant soft power. The US has lost that appeal of peace and freedom. But it [the EU] has to apply its principles (like Human Rights) when exercising power. (Author interview, March 2012)

Through its history of including smaller and poorer European countries into the EU, it has acquired credibility with respect to assisting poorer countries in economic integration (in the European market or the world economy). Several Kenyan informants called on the EU to work towards 'economic empowerment for marginalised nations. It has to go global, let them include Africa and other countries in decision making and do not just making decisions on our behalf' or expected the EU to 'help Kenya strengthen its foreign policy so as to achieve its interests. Since it has a saying in multilateral fora, it should fight for at least one seat in the UN assembly [*sic*] be given to African countries for them to have the veto' (FG questionnaire, March 2012). Overall, expectations in this category result from the EU's aspired status as a 'fairer' and more civilian global actor, guided by certain principles in support of a fair geoeconomic architecture and as a partner to African countries, in particular as regards their integration into the world economy. Indeed, Jacoby and Meunier (2010, 300) identify 'the will to manage globalization' as 'a primary driver of major European Union (EU) policies over the past 25 years'. Jacoby and Meunier's argument is primarily targeted at internal EU policies. However, it applies just as much to external expectations on the EU's global role.

Expectations in the following categories often correspond with those already discussed in the previous section. However, due to the nature of the question (EU's global role, not the EU's role in Kenya/East Africa) they tend to be more general. Additionally, here the category of *Good Governance, Peace and Security*, is a heterogeneous one encompassing expectations on the EU in the areas of good governance and human rights but also with respect to peace building and security. It clearly articulates expectations related to the second role outlined by Bretherton and Vogler (2006; see also Chapter 2), that of the EU as a promoter of normative and civilian values. In particular, European diplomats emphasise the EU's model role and civilian power orientation, arguing, for instance that the EU should,

> promote peace, development, human rights and security. (Author interview, July 2012)

> be a voice for democracy, good governance, human rights. (Author interview, November 2012)

> set clear political signals and position the EU as a polity. It should use the historical experiences of European integration. Send a message for regional integration and peace. Send of message of reconciliation and use the lessons from World War Two. (Author interview, January 2012)

> engage more in more important conflicts and matters, in the Middle East for instance. The EU can be a role model for other regions. But it needs to be more of a power politics player and show its teeth. If things don't go right, just be more brutal and leave the host country completely and completely withdraw its involvement there. (Author interview, March 2012)

In light of persistent perceptions of the EU as a moralising neo-imperial power, another member state diplomat expressed worryingly how 'we are still preaching about human rights, universal values and democracy'. However, she continued to explain that 'it is the idealist in me that wants to keep this up. If we do not, who else? China certainly not. The US, maybe, but the EU might still be more credible in this respect' (Author interview, November 2012). Also Kenyan informants viewed this category as important, arguing for instance that the EU should,

> stand for peace, equality, democracy and promotion of human rights (which would include provision of basic human needs).

> be in the forefront in addressing the increasing conflicts in many countries notably the middle East, parts of Africa for example Somalia and parts of West Africa which are bedevilled by conflicts.
>
> should put in place global mechanism to fight global terrorism and make the world a more peaceful place to live. Such terrorist groups as Al-Shabab, Al-Qaida, Boko-Haram should be wiped out. (FG questionnaire, March 2012)

In contrast to the previous section dealing with the EU's role in Kenya/East Africa, there was no pronounced difference between European and Kenyan informants in this section. Both equally acknowledged and expected the EU to act as both a guarantor of good governance and human rights but also as a security player.

Expectations of the EU's role in the three categories of *Development, Trade and Economy, Education and Capacity Building* are much less pronounced here than they were in the previous section, yet are thematically filled with similar topics. Of interest here, however, are expectations regarding the EU's role in the category *Regional integration.* For the EU's role in Kenya/East Africa, Kenyan informants expected the EU to play a role in supporting regional integration in the East African Community (EAC). For the EU's wider global role, on the other hand, the aspect of regional integration is much more relevant for European informants than it is for Kenyan informants. In fact, it has the highest number of responses from European informants, largely as a result of the long-established regional integration narrative (see Chapter 2). European diplomats frequently argued that the EU should 'present its story of peace. This is really the big message we can send out to the world. It is a role model for peace-building and regional integration, in particular in Africa' (Author interview, February 2012). Another important factor for European diplomats was also the (perceived) need to 'develop a united vision on foreign policy' (Author interview, January 2012) and to 'speak with one voice' in international relations (Author interview, July 2012). Most of the expectations on the EU in this category centred around those two issues. On the one hand, the EU is expected to serve as a model of regional integration. On the other hand, the level of regional integration, or rather the collective power of an integrated Europe, is expected to be advanced through a further pooling of resources and a stronger collective role in international relations. It has to be emphasised here that almost exclusively European informants mentioned expectations in this category.

Chapter Conclusion: Partner Without Paternalisation

I have addressed here three types of expectations: a) expectations of the EEAS as the new institution for the conduct of foreign policy; b) expectations of the EU's role in Kenya and the East African region; and c) expectations of the EU's wider geopolitical role as a global actor. With respect to expectations regarding the EEAS as the collective diplomatic institution for 'Europe', European diplomats generally shared an expectation that the EEAS should raise the EU's collective visibility and function as a coordinating mechanism to facilitate the alignment of member state foreign policies. Before the Kenyan elections of May 2013, independent agency for the EEAS was mostly foreseen in areas of added value or that were already EU competencies, such as development, trade, etc. EU diplomats acknowledged that there is probably not a 'great deal of political expectations' on the part of the member states. National diplomats expected little and rather saw the EEAS as 'one more friend to have around' on issues of diplomatic interaction with their embassies and, of course, the Kenyan government. After the elections, however, this changed substantially. The EEAS developed away from being just 'one more around' and much more towards becoming an overarching institution, pooling and 'spearheading' European diplomacy. National diplomats readily acknowledged using the EEAS to 'hide behind' when conveying potentially contentious standpoints. Relatedly, expectations for such a more pronounced political role of the EEAS rose. For interaction with cooperation partners, the establishment of the EEAS did not result in major changes. They clearly articulated that this was an 'internal EU thing'. Despite acknowledging that 'it is always easier to deal with one than with several', there are no great expectations for the EEAS on the part of the Kenyan cooperation partners.

With respect to expectations of the EU in Kenya and the East African region, the paradigm of development through trade needs to be highlighted. In Kenya, there is a strong belief in development spurred by a broadly neoliberal orientation. Trade liberalisation is seen as the way to create economic growth and development – and the promotion of trade with the EU as a key part thereof. Other approaches, for instance those associated with the paradigm of the 'developmental state', are not mentioned at all (see for instance Leftwich 1995; Yeung 2009). Whilst it could be expected that respondents in Kenya would instead advocate a neoliberal paradigm, the complete absence of any other responses is surprising. Part of the reason, next to Kenya's long-standing capitalist orientation, could be a particularly pronounced lack of confidence in state structures, resulting in the belief that the private sector is more capable of spurring 'development'. Expectations of the EU as a guardian for good governance and human rights and as a security actor also feature prominently, in particular amongst European informants, however, it is also acknowledged by many Kenyan informants, mostly those not directly associated with the government. At the same time, there are wide-spread expectations that the EU should act less patronisingly and develop a less hierarchical and more equal partnership with its (Kenyan) cooperation partners. Any (perceived) moralising or interference in domestic affairs is criticised. In particular amongst Kenyan non-governmental informants a dilemma between good governance and non-paternalisation emerged in this context. On the one hand, these informants frequently expressed their frustration with the Kenyan government's poor governance performance and high levels of corruption, expecting the EU to hold the government responsible to minimum benchmarks in this respect. On the other hand, tied aid or conditionalities are rejected as paternalising interference in domestic affairs. The dilemma is nicely summarised by an informant advocating 'non-interference in local politics but [becoming] big brother when local issues disavow international standards' (FG questionnaire, March 2012). It is clearly a fine line between principle-orientation and (perceived) paternalisation.

Conditionalities attached to budget support or other types of aid are hereby a frequent point of contention in Kenya-EU relations. In addition to the obvious argument among Europeans about the necessity of conditionalities for reasons of transparency, to avoid misappropriation of funds, or to justify to a domestic electorate that tax payers' money is being put to 'good' use, conditionalities are a key tool for *external Europeanisation*. Tying specific requirements (i.e. conditionalities) to 'cooperation' with others is a commonly used way of exercising structural power (Holden 2009) and, as such, is ideally suited to the external projection of regulated spaces of interaction as one of the EU's key geopolitical objectives (see Chapter 3). Whilst this does not necessarily limit the good governance aspect of conditionalities, it explains perceptions of neoimperialism amongst Kenyan cooperation partners and necessitates a re-evaluation of 'aid' through a critically reflexive lens that accounts for the use of development assistance as a tool of geopolitical power projection.

As regards the EU's global role, most Kenyan respondents expected the EU to function as a partner for Africa and to influence world politics towards a 'fairer' and 'more just' world order (Hill 1993; Jacoby and Meunier 2010). Economically, these responses expressed expectations that the EU facilitate the integration of African countries into the world economy to 'fair' conditions; politically, they expected the EU to function as a balancing actor between the global powers, mostly the USA and China. Closely related to this are expectations of the EU in the area of good governance and its function as a peace-builder and security provider, not only in the Horn of Africa region, but also globally. For both questions, the EU's role in Kenya and globally, the expectation of the EU to promote regional integration, is particularly pronounced amongst European informants, much less so amongst the Kenyan informants. However, in both cases it has significantly weakened compared to an earlier analysis conducted before the outbreak of the financial crisis (see Chapter 2 and Bachmann 2013a). The EU's appeal as a functioning model for regional integration, that ran consistently high before the crisis also amidst neo-imperial perceptions of the EU, is clearly losing ground in light of the financial problems in the Eurozone.

In terms of the geopolitical discourses for the EU, the results presented here show a mix of overlaps and diversions from the lead debates of civilian and normative power and as regards the three roles identified by Bretherton and Vogler. The role as a counterweight to the USA has been substantially modified. Primarily, it has decreased. European influence on world politics is perceived to be diminishing – a process that has long been described by realist critiques on Europe's role in the world (see above and Chapter 2). It has also been modified in light of perceptions of China as a major power, or rather 'partner' to African countries. And it has been reconfigured away from balancing a supposedly hegemonic USA towards the promotion of a more general 'global balance' and a 'fair' African participation in global markets and political processes. The perceptions of the EU's role as a model for regional integration persist amongst European informants, but expectations in this regard have generally become more modest in light of the financial crisis. The role for the EU as a promoter of normative and civilian standards, however, remains relatively high – albeit it is to some extent compromised by perceptions of European double moral standards as regards migration policy and (economic) neo-imperialism.

One aspect that runs coherently through expectations among European informants is the furthering of power and the strengthening of the EU's global role. European informants expect the EEAS to increase the EU's visibility and the collective influence *vis-à-vis* others. This is hardly surprising given that it is their mission to do precisely this. Yet, it also indicates a deeply rooted European will to geopolitical power – which, in the eyes of African cooperation partners and the colonial history, is suspicious and explains perceptions of continued European imperialism, justified or not. It is also an aspect that is shared amongst all geopolitical discourses on the EU's role in the world, from idealistic to realist ones. Expectations and perceptions of key EU external relations personnel are thus inextricably interlinked with lead discourses, underlining those discourses' descriptive and prescriptive nature (Nicolaïdis and Howse 2002). In particular, the civilian and normative debates have been successful in instilling an idealised version of the EU's global role in perceptions and expectations of those in charge of conducting external relations. Critical geopolitics' three facets of formal, practical and popular accounts (see Introduction) are deeply interwoven here. The lead debates describe, analyse and attempt to structure the EU's role in the world as it is practically positioned and popularly narrated. What emerges is a formal geopolitical debate formed 'by dominant intellectuals, institutions, and practitioners of statecraft' (Ó Tuathail 1996, 185) that prescribes idealised versions of Europe's global role. This debate in turn shapes how public agents, such as diplomats, perceive of their mission and the EU's role, how they conduct external relations and thereby position the EU abroad, and of course how it is narrated in popular outlets.

In terms of structure, the regulation of interaction spaces is central. Expectations related to regional integration are part of such regulation that advocates the institutionalisation and routinisation of interaction amongst different political (and economic and social) actors in a region. Also expectations related to good governance and peace have to be seen in the context of promoting the rule of law and standardised modes for the exercise of political power. In terms of contents, civilian and normative power ideas have most consistently shaped the dominant ideas and expectations of the EU's role in the world. Informants generally share the expectation of the EU to advance civilian and normative standards in world politics, ideally achieved through regulated spaces of interaction.

Chapter 8
The EU's Global Role: What Happened to the Model?

The geopolitical imagination of the EU as a 'better' global actor has long been put forward in its lead discourses (see Chapter 2) and, to some extent, also in expectations on its role as a geopolitical actor (see Chapter 7). In these debates and expectations, the alleged superiority of the European model results from a supposedly more civilian or normative orientation to international politics as opposed to military power, and the history of (West)European recovery after the World Wars during the second half of the twentieth century. This recovery has gone hand in hand with the process of regional integration, first in Western Europe and after the end of the Cold War also in Eastern Europe; and thereby with the creation of a complex system of regulated spaces of interaction (see Chapter 3). In many ways, the regional integration processes are thereby credited – whether justified or not – with delivering peace and relative prosperity to a continent historically prone to internal war. For Europeans, the regional integration process, fed by civilian and normative power ideas, thus appeared indeed often as a 'better' geopolitical model.

For outsiders, in particular for those subjected to European colonialism at a time, however, this supposed superiority of the European model is not immediately evident. The benefits delivered to Europeans as part of the integration process, for instance free movement of people and goods but also more general aspects of peace, prosperity, exchange opportunities, etc., are of no benefit to citizens of African countries. On the contrary, through the creation of one large European polity, the negotiation position of African countries *vis-à-vis* 'Europe' has significantly weakened. Clearly, the example of peace and prosperity is appealing, yet the 'success' of the model in Europe does not imply that it is replicable elsewhere. Moreover, in terms of the appeal of prosperity, the situation has significantly changed with the outbreak of the financial crisis in Europe. The situation of economic downturn in many European countries has been furthermore in contrast to a seemingly ever-growing China. If the European model is thus indeed superior, why is it apparently incapable of resolving the current problems in Europe? In addition, if 'Europe' is supposed to be a superior cooperation partner to African countries, why have four decades of European development aid failed to produce visible results when the more recent Chinese engagement produces visible 'development' *en masse*, mainly in the form of large infrastructure projects? Also, where is Europe's normative orientation and insistence on human rights in its bordering and migration policies? Clearly, these questions are too complex to be answered swiftly, yet they underline prevalent doubts about the superiority, and the normative morality, of the European model. Any attempt at approaching the EU's global role, and how the EU is seen as a geopolitical actor, inevitably faces such questions and doubts about the superiority of the European model that are so frequently postulated in dominant discourses.

This chapter offers such an attempt. It addresses informants' perceptions of the EU's wider geopolitical role as a global and development actor. Whilst the previous chapters were predominantly concerned with specific issues regarding the EU's role in Kenya or as a global actor, such as *Europeanisation abroad* (Chapter 5), *external Europeanisation* (Chapter 6) or expectations on the EU's role in Kenya and as a global actor (Chapter 7), this chapter engages with informants' general perceptions of the EU. It analyses responses to the question of what they primarily associate with the EU and its geopolitical role as well as to the questions of what they regard as the most positive and most negative aspects of the EU.[1] In particular the responses to questions asking for aspects perceived as particularly positive or negative reveal stark differences depending on the informants' background (i.e. Kenyan or European). I also reflect here on the wider perceptions illustrated in the second part of the book in light of the Eurocrisis and other major geoeconomic and geopolitical transformations, such as the increasing Chinese engagement in Africa.

What Kind of a Global Actor is the EU?

> The EU is a block of 27 countries and associate members with common internal and external policies to promote interest of its members. It is a formidable union that gave exemplary history to other continents that divided we fall, united we stand. But recently its Euro crisis seems to be a negative lesson to others. (FG questionnaire, March 2012)

One of the key attempts to classify the EU's role in the world is that put forward by Bretherton and Vogler, who outline three roles of the EU as a model for regional integration, as a promoter of normative and civilian values, and as a geopolitical counterweight to the USA (see Chapter 2 and Bretherton and Vogler 2006, 56–7). Bretherton and Vogler's approach still offers a useful differentiation to enable a rough structuring of different aspects of the EU's role in the world. However, it falls short of accounting for how the EU is seen by its external cooperation partners and of including more recent major geoeconomic and geopolitical

1 As for the graphs in Chapter 7, the graphs presented here combine responses given during the interviews and the focus group discussions (including written responses in the form of the preceding short questionnaire). Taken together, the number of (Kenyan) cooperation partners included is thus higher than those of European diplomats (88 to 36), causing the mismatch between the upper (lighter) and the lower (darker) parts of the graphs (see also Table MR.1 and explanations in Chapters 4 and 7).

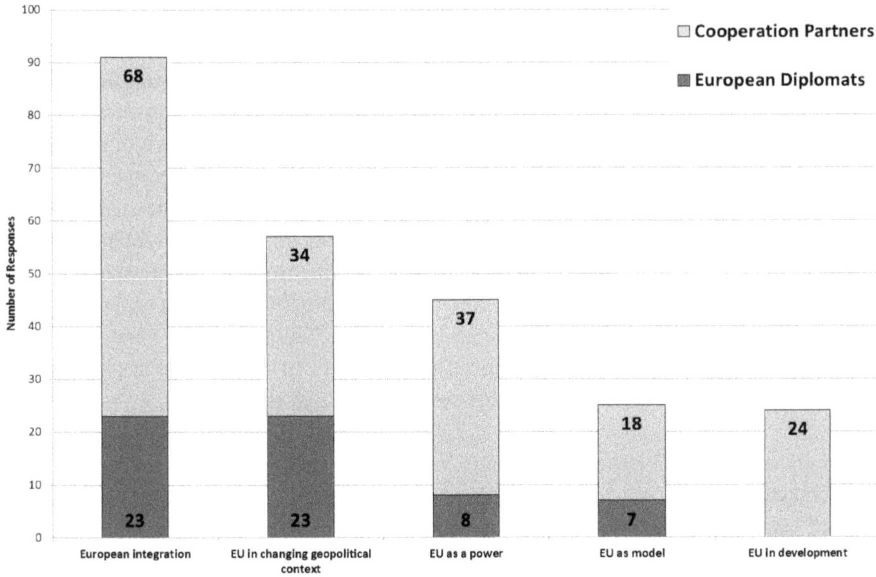

Figure 8.1 Perceptions of the EU as a geopolitical actor

transformations. Based on previous research on external perceptions of the EU, I have elsewhere laid out an alternative differentiation of the EU as an appealing civilian geopolitical model and as a resented geoeconomic power (Bachmann 2013a). This differentiation emerged from empirical research conducted mostly in 2007 and 2008, namely before the outbreak of the financial crisis in the Eurozone. The current work presented here, however, does take into account the effects of the crisis on how the EU is seen from the outside. Nevertheless, despite the doubts regarding the functioning of the EU's model that emerged in the wake of the crisis, regional integration – very broadly defined – is still what the EU and its geopolitical role is primarily associated with (see Figure 8.1).

Regional Integration: The Inherently 'European Thing'

'The inherently 'European thing' is regional integration' an official from the Directorate-General Development in Brussels stated during an interview in 2008. Notwithstanding recent cracks in the story of successful regional integration, this assessment of what 'Europe' is essentially about steadily persists. Figure 8.1 shows how also more recently, the aspect of regional integration is the one which is most frequently mentioned with respect to the EU's geopolitical role. Responses in this category refer to a broad variety of aspects roughly related to the process of European integration. This often includes aspects such as the common market, the Euro, regionalism, trade, but also more general responses such as 'common policies, interests and goals', 'European countries coming together' (FG questionnaire, March 2012). Responses cover a range of positive attitudes, referring to successful regional integration in the interest of European people and countries, and more negative viewpoints describing the integration process as a pooling of power to dominate other countries and regions. This range of responses is to some extent also reflected in the other categories (model and power).

In terms of backgrounds of informants, it is unsurprising that European informants mostly pointed to the positive aspects of regional integration whilst Kenyan informants also tended to critically point to some of the more negative aspects, often referring to an amassing of European power through the integration process – similar to the critique outlined by Johan Galtung in the 1970s (Galtung 1973). Foregrounding the positive aspects with respect to Europe's history of overcoming war through the integration process, a member state diplomat argued that the

> only thing the EU can bring to Kenya and to East African countries is its story and the fact that we decided as a bunch of six countries to revise the course of history. My generation is the first generation of Europeans who hasn't been fighting each other. It's just an example of how a bunch of us got together and manage to reverse the course of history and that's the only thing we can actually bring to anywhere else: […] the experience of a story that somehow, against all odds, we moved forward (FG1, March 2012)

The statement is indicative of many others that frequently point to the appeal of Europe's story of overcoming war and creating an area of relative peace and prosperity in the aftermath of World War Two. Such perceptions were part of the responses from Kenyan and European informants alike. During a focus group discussion, for instance, a Kenyan informant responded that, to her, the EU is 'the solidarity with which Europeans are thinking of coming together' (FG3, March 2012). More neutrally, another Kenyan respondent described the EU as a 'union of countries in Europe with common trade agreements, free movement between member states, common currency, common monetary policy, conflict management among member states, environmental issues' (FG questionnaire, March 2012). In the focus groups, informants revealed a mix of attitudes with a general sense of admiration for managing the processes of integration after centuries of warfare; often coexisting with a resentment of the pooling of power on a European level to the disadvantage and exclusion of Africa(ns). During a focus group discussion and in response to a projected PowerPoint slide asking 'What is the 'EU'rope to you?', a Kenyan informant responded,

> [i]t is a rope that is tied around a union of European countries that have agreed to work together. It is spreading its tentacles to other countries of the world. It is the beginning of one-world government. It is a structure that brings the European Union together, but that can also break up anytime (FG2, March 2012)

Similarly inspired by the orthographic style of writing 'EU'rope, another informant responded to the same question at a different focus group discussion, stating that

> E is European, U is Union, 'rope' is just used to tie themselves together with a rope. Nobody can enter. This way you can see why going to Europe is difficult, even though some have gone through Libya, but it is difficult because European countries tie themselves together with a rope. (FG3, March 2012)

The metaphor of the rope is an interesting one – one that is highly expressive of the European approach of integrating internally, including the freedom of movement, whilst shutting itself off from the outside. Often referred to by the term 'Fortress Europe', the rope metaphor also captures the biopolitically problematic and discriminatory approach of European migration policy whereby the right to migration is not only granted but encouraged to human beings with European passports whilst simultaneously and systematically denied to others – in particular Africans. In light of the European rhetoric focused around human rights, non-discrimination, and equal rights and treatment to all human beings (for instance with respect to gender or LGBT issues), African perceptions of European double-morality are a logical result – and, of course, severely undermine Europe's moral authority in human rights questions.

Despite the frequent emphasis on its integrated character, European informants in particular also pointed to the transitional phase of the EU's current geopolitical role. A member state diplomat, for instance, explained how the 'member states are reluctant to give the EEAS high profile persons because they don't want it to become overly dominant. Member states need to have the courage to equip the EU/EEAS with the people and the mandate to have a real impact. At the moment they are often afraid of doing so' (Author interview, November 2012). At the same time European informants acknowledged that the EEAS is still a relatively recent attempt at establishing a global political role for the EU, arguing that 'it just takes time. It has always been like this: EU treaties were ahead of their time and reality then adapted after some time. It will be like that with the Lisbon Treaty and the EU's global political role, too' (Author interview with member state diplomat, February 2012).

More generally, most responses in this category viewed the EU in its function as an integrated bloc, sometimes highlighting more positive, sometimes more negative aspects about 'moving together'. What united the responses, however, was a general take on the EU's integrated character in positioning it as a strong geopolitical actor in the wider field of international relations. Responses in the next category reflect more critical attitudes as regards the EU's geopolitical role in light of major geopolitical and geoeconomic transformations, such as the Eurocrisis or the growing role of China in Africa.

The EU in a Changing Geopolitical context

> I appreciate the long history that Europe has had in coming together, but for me, I see it as a means for survival in the changing international system. Polarity is changing, China is coming up, the international system changing very rapidly. I see it largely as a means of the European Union, of the European countries wanting to remain relevant and remain a force, a hegemon if you like, in the international system. When they come together they wield more of everything, more power, more resources, more markets. (FG3, March 2012)

The high number of responses in this category is indicative of relatively recent changes in (perceptions of) the EU's geopolitical role. Whilst external perceptions in an earlier research project (with fieldwork undertaken in 2007/2008) roughly responded to the three roles outlined by Bretherton and Vogler plus perceptions of the EU as a geoeconomic power (often in a neoimperial context), this category reflects recent changes in such perceptions. The two dominant strands of an appealing civilian geopolitical model and a resented geoconomic power do not completely vanish. However, both are to some extent dimmed in the wider context of a generally perceived weakening of the EU's global role. In a way, responses in this category reflect a waning of European influence in the world, whereby both the appeal of its model as well as the resentment against its power is diminishing. This diminishing global role is vaguely fixed on two key lines of arguments, the Eurocrisis and increasing Chinese influence in Africa (see also Keukeleire and Hooijmaaijers 2014).

With respect to the crisis in the Eurozone, in particular European informants clearly pointed to its effects on the perceptions of the EU's international role, arguing, for instance, that the 'perception of the EU as a globally important economic power is just being qualified through the crisis. It gets less!' (Author interview with EU diplomat, March 2012). Keenly aware that 'our economic clout has clearly suffered' a member state diplomat argued that much of the EU's global role in the future will depend on 'how well we are able to pull ourselves together' (Author interview, March 2012). More critical of the EU's future global role in light of a growing Euroscepticism in some countries, another member state diplomat argued that

> there were times where you could see the EU as the big geopolitical project in a sense of bringing the European continent together. The Eurocrisis has revealed that in most countries most people do not want a political union. This is tied to also the EU's problems in accountability and a democratic deficit. In many ways a better cooperation is the only way to go. The entire project must not fall apart, but that does not mean it is to become a political union. So, it is hard to see a geopolitical role for the EU coming. (Author interview, November 2012)

Even more drastically, another member state diplomat asked rhetorically if the EU is needed at all given its inability to deal with the economic crisis. Responding to the question about the EU's global role, the informant then continued, 'I do not see a global role for the EU. The EU is not a partner for the US, China, etc. They prefer bilateral partners' (Author interview, March 2012). However, not only in the eyes of European diplomats was the crisis in the Eurozone seen to severely impact on the EU's international reputation and actorness. Kenyan informants also doubted both the European 'model' of regional integration as well as the EU's capability to actually play a leading international role as a global and/or development actor. Not without sarcasm, an informant remarked how 'the west always warns of Chinese involvement in Africa, partly also because of human rights problems. Now it is the West that turns to China for support in its own problems, like the crisis of the Euro' (RD notes, February 2012). Also during the focus group discussions on the EU's global role and with reference to Bretherton and Vogler's (2006) third role of the EU as a counterweight to the USA, participants reflected widely on the EU's changing position in this context, arguing that the

> EU tends to cooperate more with the US. It is more of the same, not so much different, they often stick together. It is not so much a counterweight to the US. They [EU and USA] are very close due to historical ties. If we look at voting patterns in UN or so, EU countries are not counterweighing the USA. China is doing it much more. Also economically, China is much more visible in Africa, so the third role [as counterweight to the USA] is really a statement for China. (FG2, March 2012)

or that

> I don't think that the EU is in a position to play the role of a counterweight role now. I think it could be in a very good position to play that role in future but now, because of the economic crisis, the EU has a bit of weakness. But to be able to play the role of a counterweight, you must have some weight to spread around. Right now they might be more concerned with China. (FG3, March 2012)

Other informants highlighted how a combination of factors, paring the uncertain economic situation in the Eurozone with European aid conditionalities and the Chinese alternative of non-interference, essentially forced African countries to 'Look East'. Focus group participants argued that because of the 'decline of specific European countries' through the Eurocrisis, the EU will be primarily concerned with its internal problems and European support to Kenya will be unpredictable. Also in light of the 'stringent conditions put by EU countries', the informant continued, this 'leaves us with no option but also to look East, where it is easier to get the access and all those things without much conditions' (FG1, March 2012). An official at the Kenyan Ministry of Foreign Affairs nicely remarked how 'Europe still is a significant partner for us. But it needs to realize that the Africa of 2012 is not the Africa of the 50s, 60s or 80s' (Author interview, December 2012). Neither is Europe, nor the world around it.

EU as Power

> EU means a representation of member states of the European countries. First things that come to mind are 1) conglomeration of super powerful states in the world; 2) states pursuing their interests; and 3) powerful states in the world dictating to other "smaller" states. (FG questionnaire, March 2012)

Responses in this category are somewhat contradictory to those in the previous one, since they emphasise Europe's power position, whilst the previous one largely emphasised the decline of such a position of power. For Kenyan informants, the EU's role as a power, mostly seen in geoeconomic and neo-imperial terms, is even the second most frequently mentioned aspect (after the EU in terms of integration). EU economic power is thereby often, yet not exclusively, seen as negative and oppressive for African countries. Pointing to the EU's role in 'stabilizing the world economy', an official of the Kenyan Office of the President, argued that 'in a global village you cannot talk about the global economy without talking about Europe' (Author interview, March 2012). This roughly articulates with longer-established roles of the EU as interlocutor in global South-North relations (Hill 1993), as a manager of globalisation (Jacoby and Meunier 2010; Mügge 2014) and respective expectations (see Chapters 2 and 7). During the focus group discussions, the EU's role as a (neoimperial) power was also intensely discussed, often in geoeconomic terms, however also in light of the EU's support for the cases against Uhuru Kenyatta and William Ruto at the ICC (see Chapter 6). The ICC was thereby portrayed as a neocolonial institution 'formed by powerful countries to advance their interest' and 'for putting Africans on trial' (FG1, March 2012). Not only with respect to the ICC, but even more so economically, the EU was seen as a 'means of some European countries [that] have joined together to impose their policies to the underdeveloped countries of course for their own gain' (FG questionnaire, March 2012). Such perceptions of the economic side of the EU's power largely prevailed in this category. Informants viewed the EU as an 'instrument to dominate less developed countries and also an instrument of imperialism' or as a 'group of European countries that have come together so that they can have the power and might to influence the world' (FG questionnaire, March 2012). European diplomats are well aware of the persistence of such neoimperial perceptions. After one of the focus groups discussions, an EU diplomat laid out how he

> can also see where such perceptions come from. Some EU-countries, in particular the former colonial powers, still have very strong national interests in Africa and seek to pursue them and influence the situation in African countries to their benefit. This is of course noticed here and hits resentment. On the other hand, I think the enlargement of the EU was extremely good in this respect. The new countries in Central and Eastern Europe (and also the smaller Western European countries) are a lot more credible in this respect and can bring back some anti-imperial credibility to the EU as a whole. (Author interview, March 2012)

The 2004 'Big Bang' enlargement has elsewhere been identified as an important factor in dispelling imperial perceptions of the EU – but also as an example of how the EU assisted smaller and economically less powerful countries to integrate into the EU and the world economy more generally (Bachmann 2009, 73). Nevertheless, in particular in light of the Kenyan elections and the ICC cases, neoimperial perceptions of the EU continue to loom large among African informants. Yet, even though there was a general criticism of a (perceived) oppressive relationship, some informants expressed an understanding view with respect to the exercising of EU power, arguing that

> [o]f course as an African, I wouldn't want African states to be oppressed, but from an academic standpoint I understand completely why it is important [for the EU] to strengthen its power side. Also from a legal perspective, when you are negotiating a contract the main aim is to take advantage of the other party. If the other party is asleep, you know, I mean too bad for them. So, as an African I would want to see a reduction of European power [in African-European relations], but if I was in the same position as the EU, meaning that Africa was stronger than the EU, I would not abandon the strategic advantage of the stronger. (FG3, March 2012)

This view is both problematically neorealist and indicative of a demand for the self-responsibility of African leaders in their negotiations with the EU. An oppressive relationship is primarily seen as resulting from the lack of negotiation skills of African leaders and not of the EU's inherent neoimperial character. Whilst this view is clearly problematically ignorant of larger geoeconomic power structures and measures that exert influence over marginalised countries, it does hint at African leaders' responsibility towards their countries in light of a wider atmosphere or of coating internal mismanagement, poor governance and corruption with such larger macrolevel power structures. Nevertheless in general, most responses in this category viewed the EU as a one of the dominant powers in the world; playing an influential geopolitical and geoeconomic role, focused on the promotion of its self-interest. This was often associated with imperialism and the marginalisation and domination of African countries.

EU as Model

> The EU is a supranational organisation, is the embodiment of my dream for Africa – an "African Union" – whereby all African countries come together, under an umbrella body governed by one currency, multiple streams of economic, political as well as diplomatic engagement.

> The EU is a union of countries in Europe with common trade agreements, free movement between member states, common currency, common monetary policy,

> conflict management among member states, environmental issues. (Responses from FG questionnaire, March 2012)

Responses in this category differ significantly from those in the previous one, in that they recognise previously identified roles of the EU as a civilian geopolitical model (see Chapters 2 and 7; also Bretherton and Vogler 2006; Bachmann 2013a). Instead of criticizing the exercise of 'oppressive' European power in a hierarchical relationship with African countries, responses here emphasise the EU's potential to act as a model. Despite the doubts arising about regional integration illustrated in the section above, the history and presence of European integration continues to have a substantial appeal whereby the EU remains viewed as a model and promoter of integration efforts abroad (Jetschke and Murray 2012; Lenz 2012; Hulse 2014). Other, yet related, aspects in which the EU is seen as a model are in areas of human rights, good governance and civilian standards, as well as with respect to peace, economic and social liberties. Most responses in this category contained strongly positive connotations about the EU, often highlighting its civilian orientation.

However, with respect to the EU's role as a model for regional integration, there is a keen awareness amongst the EU's cooperation partners to not 'build a European Union in East Africa but to build an East African Community' (FG1, March 2012). Informants highlighted the need for careful attention to positive *and* negative aspects of the European experience of integration and the aligned approach of not just replicating the process of European integration. In particular in light of the Eurocrisis, African informants increasingly emphasised an approach of learning from both good and bad experiences in order to create a *locally adapted* integration process, specifically pointing to the need to

> distinguish between the lessons to be learnt. Not all lessons are positive, some of them we learn them so that we are not making the same mistakes or repeating the same problems, for instance the issue of the debt crisis. There is something that we can learn. Not necessarily to implement it, but be able to avoid it. (FG1, March 2012)

> Similarly, during a large focus group discussion, a group of participants expressed a general appreciation of the EU's model of regional integration but equally cautioned against negative developments associated with the financial crisis:

> We see Europe as an example of what integration should be but we are still waiting on the verdict of that because of what is happening right now with the Eurocrisis and so. We think it is a good example for the East Africa Community. But maybe we shouldn't do a photocopy of what Europe is right now and wait for the process and see what happens, so that we can then copy the end result. We want to learn from your mistakes as well and it will be interesting to see how you resolve your current issues so that we can hopefully not fall into the same traps. (FG3, March 2012)

Nevertheless, most responses in this category were largely positive, both among European as well as African informants. European informants often pointed to the history of a continent growing together after the wars in the first half of the twentieth century and, with the 'Big Bang' enlargement, also after the Cold War (Author interviews 2012). A central aspect of the integration narrative, which is at the root of the EU's model character, is its civilian orientation and (the perceptions of) its commitment to peace *vis-à-vis* other major geopolitical and geoeconomic powers. A member state diplomat, for instance, underlined the EU's 'universal values of a soft power. This is a fundamental difference, for example to China. We act *not only* in our self-interest!' (Author interview, March 2012). More generally, however, the perceptions of the EU given in this category roughly correspond to the expectations of the EU as a promoter of regional integration elsewhere (see Chapter 7), and also to the roles ascribed to the EU in previous research projects (Bretherton and Vogler 2006; Fioramonti and Poletti 2008; Bachmann 2009, 2013a). Yet, in light of the Eurocrisis, the EU's model character has clearly suffered and its role as a promoter has become more subdued in relation to other aspects (see Figure 2.1 and 7.2).

EU in Development

> The European Union is a union of about 27 countries that have come together as an international organisation with a common identity for a common goal. First things that have come to mind: Oversees development assistance, humanitarian aid, Euro currency. (FG questionnaire, March 2012)

Responses in the last category refer to the EU's role in international development and humanitarian aid. None of these responses came from European informants. This can be explained by the nature of the question which asks about aspects the informants primarily associate the EU with – it is not surprising that European informants tend to mention internal aspects first. For African informants, on the other hand, 'the EC is a very important organisation in development' (Author interviews, March 2012). It is no coincidence that for African informants this category is more important than seeing the EU as model, given the EU's scope of aid as the largest donor and its long-standing engagement in development cooperation and humanitarian aid. Focus group discussants described the EU as 'the largest donor in the world. EU facilitates human development in most parts of the world through grants and contracts. EU also facilitates research related projects', and even argued that the EU's 'major mandate is to bring European aid closer to developing countries' (FG questionnaire, March 2012). More generally, despite the extensive Chinese engagement in Kenya and the wider African continent, European development assistance continues to play a major role, also given that in contrast to Chinese assistance most European cooperation is channelled as grants not as loans. The high number of responses from African informants in this category also corresponds to the respective expectations of the

EU as a key development actor (see Figures 7.1 and 7.2) and of the EU primarily seen in a development function as opposed to a political/diplomatic actor (see Chapter 6).

The EU's Fame and Shame

The section above has illustrated a range of informants' perceptions of and views on the EU as a geopolitical actor. Some of the roles ascribed to the EU were seen in a positive light, others in a negative one, and yet others were relatively neutral descriptions. In addition to inquiring about what informants primarily associate with the EU, I asked informants about their spontaneous opinion regarding what they consider the three most positive and three most negative aspects of the EU. The questions were kept open and inquired in very general terms about good and bad aspects that informants associate with the EU. There was no intention to focus on African-European relations, however, given that the preceding interviews touched significantly on this relationship, it was also reflected in many of the responses in this category. The quantified results of these responses are shown in Figures 8.2 and 8.3.

On the positive side, responses were split between issues concerning the nature of the EU as such and those concerning its relations with Africa. Most responses (both by African and European informants) were given in the category *Regional Integration and Civilian Orientation*. This is consistent with dominant geopolitical narratives of the EU (see Chapter 2), with the most dominant perception of the EU's geopolitical role in general (see Figure 8.1), the results my own previous research projects (Bachmann 2009, 2013a, 2013b) as well as with those of other related research (Lucarelli and Fioramonti 2010; Börzel and Risse 2012b; Holland and Chaban 2014). It also indicates that the EU's appeal as a model of regional integration and civilian power (see Bachmann and Sidaway 2009b; Manners 2010) has not completely vanished. Beyond the regional integration aspects, responses in this category often referred to the EU's broader 'civilian' orientation, its preference for multilateralism, as well as a (perceived) difference from other powers such as the USA and China. Responses also included issues around the free movement of people and goods in the union, the EU's commitment to democracy and environmental issues, as well as the common market and the Euro.

The overarching civilian power orientation is also reflected in many of the aspects mentioned in the other categories. Responses in the category of *Development, Education, Good Governance and Human Rights in Africa*, for instance, referred to perceptions of the EU as a supporter of African countries with respect to internal 'development'. This was often mentioned in line with democracy promotion, the support of good governance and human rights, education and training schemes, humanitarian assistance, peace, infrastructure and agriculture, etc. Whilst for European informants, this category only played

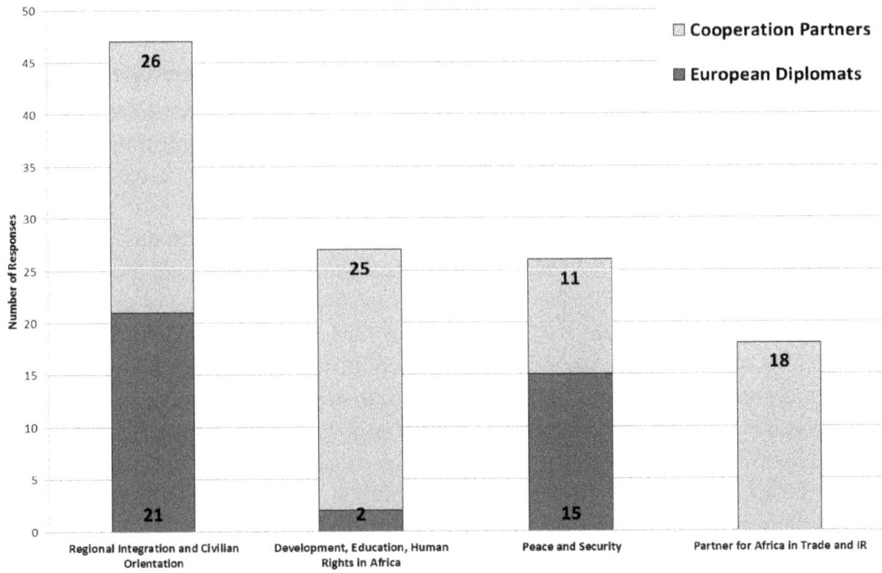

Figure 8.2 Best things about the EU

Textbox 8.1 Best things about the EU (examples of responses)

Free movement of people, ideas, goods

It's not all about individual advantage and opportunism

Advocates for global cooperation

It has a very good dispute resolution system

Overcoming of the Cold War division

The historical peace project of European integration

Their support to maintain peace and avoidance of using their military power

EU has contributed to world peace and somehow stabilized the world economy

Peace, regional growth and development

Pushing for democratization and funding development projects in Africa

Its willingness to open its economies to third world countries so they can export goods such as flowers

a minor role, for African informants it was almost as important as the integration and civilian aspects of the first category.

The EU's broadly perceived peace-orientation and its role as a security actor were at the centre of responses in the category of *Peace and Security*. Informants described the EU as strongly committed to peace and promoting peaceful structures for the settlement of conflicts, but also as a security actor in peace-building and peace-keeping missions. In particular for European informants this constituted an important category, whereby most European informants pointed to EU's peace-orientation and African informants frequently mentioned the EU's role as a security actor, in particular in the Horn of Africa. Responses in the smallest category, *Partner for Africa in Trade and International Relations*, referred to perceptions of the EU as a partner to African countries that promotes international trade and supports African interests on a global stage. All of the responses here were given by African informants. It is hardly surprising that European informants tend to identify more Eurocentric aspects as the most positive (and most negative) about the EU – in contrast to African informants. The EU's role as partner for Africa, thereby does not play a role in their views to this question. What is slightly surprising, however, is that in light of the neo-imperial perceptions of the EU, it is still – at least to some extent and more so than for instance the USA or China – seen as a partner that does 'not only act in [its] self-interest', as a member state diplomat put it.

On the negative side, responses were equally split between African and European informants, whereby most of the former pointed to a neoimperial character of African-European relations and most of the latter addressed internal EU-issues. For African informants, the most negative aspects were tied to a perceived hierarchical and discriminatory European behaviour towards Africa(ns) – both in terms of economic and political relations but also in terms of daily and personal interaction. Responses in the category *Discrimination and Xenophobia* criticised the EU for double standards on human rights, for 'blackmailing' Africa, and for its visa and immigration policies. Most of these responses were voiced in the context of discriminatory practices or attitudes and behaviour towards Africans in both legal frameworks and the European public, for instance European migration regimes or experiences of racist interaction. No European informant gave a response in this category.

In sharp contrast, most of the responses by European informants were in the category *Bureaucracy, Costs and Democratic Deficit*. Unsurprisingly European informants were more concerned with inner-European issues, in particular the inefficient management of funds and bureaucratic overregulation through Brussels' institutions. Some informants also pointed to the lack of democratic accountability of Brussels' 'Eurocrats' and their distance from European populations. Responses from African informants in this category primarily pointed to the heavily bureaucratised and stringent procedures when cooperating with the EU in joint projects or with respect to how funds were allocated and channelled.

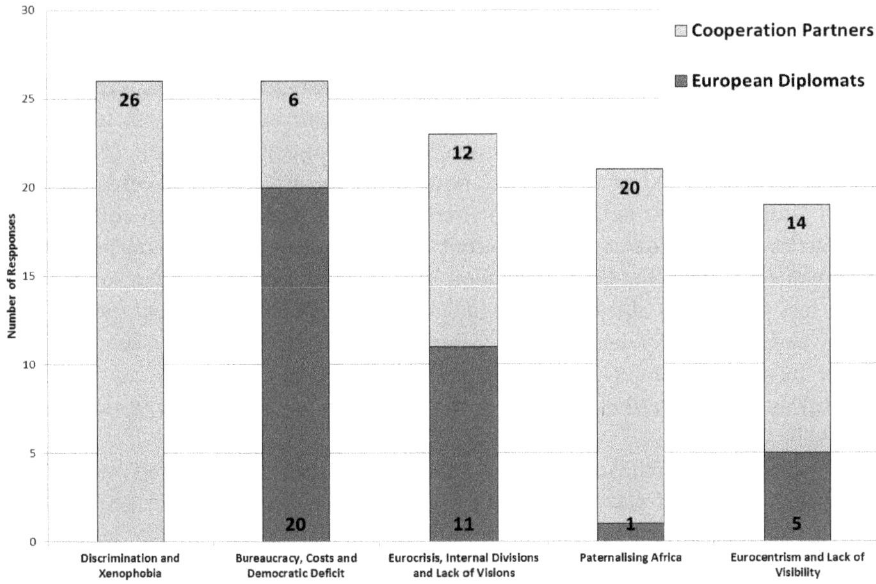

Figure 8.3 Worst things about the EU

Textbox 8.2 Worst things about the EU (examples of responses)

Denying visas for Africans, which means they still see us like slaves

Mistreatment of Africans who are in Europe

EU is neocolonialist in its relations in Kenya

Imposing their policies on weaker states

Quit with the carrot and stick way of management in Africa

Sometimes it appears as a monster that cannot be controlled

The EU needs a lot of money; it is an exorbitant bureaucracy with no effect

It should listen to its citizens

Lacking solidarity within the EU or actually in the MS (in particular as now in times of crisis)

Failure to equally empower its members

No vision, no orientation, no target

People are inward-looking

EU is passive when it comes to dealing with world conflict

The next category, *Eurocrisis, Internal Divisions and Lack of Visions*, is an interesting one with a roughly equal share of responses from African and European informants. Responses from the EU's cooperation partners were mostly concerned with internal divisions within the EU contributing towards letting the EU drift into an unreliable and unpredictable cooperation partner. Informants voiced a certain level of uncertainty both about the EU's general orientation in terms of development policy and its commitment to support African countries as well as the EU's future capability to function as a development partner in light of diminishing financial solvency induced by the Eurocrisis. European informants were less concerned about a future lack of commitment to development policy, however they equally lamented increasing internal divisions between member states, increasingly emerging in the context of the crisis, as well as the lack of a general future orientation as to where the European (dis)integration process is or should be going.

Responses in the category of *Paternalising Africa* refer to perceived inequalities in African-European relations, primarily in larger geopolitical and geoeonomic terms. This includes the imposition of policies and of specific spaces of interaction and modes of policy conduct in development policy and economic cooperation, for instance through aid conditionalities. It also includes the ignorance of local solutions, approaches and customs, inference in domestic affairs and the active promotion of continued hierarchies and dependency. Whilst for African informants, this constituted the category with the second highest number of responses, only one of the European informants identified such hierarchies as problematic. This is surprising given that European informants also tend to be aware of the persistence of neoimperial perceptions of the EU in Africa. Yet, mainly concerned with inner-European issues, they tend not to identify such perceptions as a major problem.

Related to European informants' prime concern with inner-European issues, responses in the category *Eurocentrism and Lack of Visibility* criticised the primarily inward-looking orientation of the EU, often associated with a perceived lack of European interest in Africa. African and European informants alike lamented a European passivity and reluctance to actively contribute to solutions of problems outside Europe. Some informants also criticised that the benefits of the European integration process are only for Europeans and European countries whilst simultaneously excluding non-members and displaying an indifference towards the rest of the world. Similarly, respondents also criticised the EU's subdued role as a major voice in development and security in Africa, often in contrast to a much more visible Chinese and American role in these aspects.

Chapter Conclusion: Positive and Negative Lessons from the European 'Model'

Unlike the previous chapters, this chapter is much less concerned with immediate experiences of the informants' work in or with EU institutions, let alone the European External Action Service. It inquires into informants' perceptions of what the EU is and primarily means to them, and, furthermore, which aspects of the EU they regard as positive and which ones as negative. The chapter is thus not about informants' expectations of what they want the EU to be or what they expect from the EU, but very generally on how they primarily see the EU as a geopolitical actor.

A clear trend that emerges is that the EU is still primarily seen in terms of regional integration. Pared with its general civilian power orientation, this is mostly perceived as positive, however with an increasing caution in light of the financial crisis in the Eurozone. It is no surprise that regional integration emerges at the top of associations with the EU, as this is how the EU presents itself and is articulated in both official and academic discourse (Chapter 2) and also how the EU has long been seen by its external cooperation partners. What we can observe over the past five years and in light of the Eurocrisis, however, is a break in external perceptions as regards the appeal of the EU's model and expectations to promote that model abroad. On the one hand, general perceptions of the EU as both a regionally integrated and more civilian geopolitical actor as well as of the EU as a neoimperial geoeconomic power (Bachmann 2013a) have remained fairly consistent. On the other hand, this does not automatically entail any more expectations that the integration model should also be projected abroad. Before the crisis, the history of European integration in the second half of the twentieth century, associated with the creation of peace and relative prosperity, was largely seen as a success story, tied to the regional integration narrative. Regional integration appeared to be the solution for creating peace and prosperity (Fioramonti and Poletti 2008; Bachmann 2009), and the EU was widely expected to promote regional integration elsewhere. The prescriptive character in the EU's lead discourses thereby not only applied to internal vision, but also guided external projection (see Chapter 2). In the context of the Eurocrisis, this narrative has retained some appeal for its civilian orientation and for creating peace, however, it has lost appeal as a model for creating prosperity. Simultaneously, the rapidity of Chinese economic development over the last two decades, and its increasing engagement in Africa, offers a different model for economic development, centred around state-led autocratic capitalism. In particular for political elites, this model is highly appealing, not only for its quick and visible gains in terms of 'development' and infrastructure, but also because it ascribes much power and little accountability to the political elites and limits 'concerns' with good governance and human rights. In contrast, the European 'model' is specifically centred around such aspects, yet due to the crisis, it has failed to deliver what it was previously admired for: growth, economic development and prosperity. The crisis, and its social and economic impact in many (Southern) European countries, has revealed the pitfalls of a large, macroeconomically inert polity comprised of different economies with very different capabilities and interests. In addition, a diminishing sense of 'togetherness' in European populations, fuelled by a highly worrying rise in nationalism across Europe, has painfully revealed the thin ground on which regionally integrated solidarity is built.

It is in such a wider context that the regional integration narrative has lost some, but not all, of its previous glamour. It is still recognised as a peace project and admired for the internal freedoms it has generated on the European continent. Yet it is also seen more critically in light of the economic crisis for its inability to resolve it – possibly even for causing it in the first place. The crisis has thus produced a more sober take on regional integration and its potential as the EU's prime geopolitical export. Instead of being seen as a one-size-fits-all solution for the creation of peace and prosperity and of integration for its own sake, regional integration is now regarded more pragmatically as one tool amongst others towards achieving certain political and economic objectives. For some of these objectives it continues to be seen as a suitable tool, for others less so – or at least existing in fierce competition with other tools. Through the crisis a more differentiated view of regional integration has emerged – one that continues to see the EU as a model, but a model that teaches not only positive but also negative lessons in terms of integration steps to avoid. It has also underlined the importance of not blindly copying any model, but of carefully considering positive and negative lessons as much as local contexts – of 'not trying to build a European Union in East Africa, but of building an East African Community', as a participant in a focus group discussion put it.

Whilst wider geopolitical and geoeconomic changes, such as the Eurocrisis and the growing influence of the BRICS-countries have clearly impacted on informants' perceptions of the EU's global influence, other perceptions have remained fairly consistent. The EU continues to be seen as an appealing civilian geopolitical model committed to good governance, human rights, and peace. But equally it continues to be seen as a neoimperial geoeconomic power (Bachmann 2013a). Also in terms of Bretherton and Voglers' (2006) three roles of the EU as a model for regional integration, as a promoter of normative and civilian values, and as a geopolitical counterweight to the USA, perceptions have changed. For the first role, the EU continues to be seen as a model for regional integration, but that narrative has lost (part of) its appeal and model character in itself. The second role clearly retains perceptions of the EU's civilian and normative orientation. Yet both the first and the second role lose the automatism of external projection. The last role, the EU as a counterweight to the USA, has been significantly modified (see also Chapter 7). In light of the Eurocrisis and a growing Chinese geopolitical and geoeconomic role, the EU appears too weak to be a significant geopolitical counterweight to any major global power. On the positive side of perceptions, and in addition to its civilian power orientation, the EU continues to be seen as a development actor and as a 'partner' for African countries, mainly with respect to their integration in the global economy. Moreover, and surprisingly, as an actor in peace and security, the EU is seen in a largely positive light. On the negative side, the overarching perception of the EU is that of a major geoeconomic power pursuing its self-interest, demonstrating paternalising, disadvantageous and discriminatory politics (and people) towards Africa(ns). For European informants, the key problems were found less in the EU's hierarchical relations with Africa, but internally, such as a bureaucratic inertia, the Eurocrisis or a lack of a general orientation regarding the future of the (dis)integration process.

The perceptions presented here capture, naturally, only a small snapshot of perceptions about the EU's role in the world. Nevertheless, they are indicative of the complexity of the external relations of this heterogeneous polity. There is no one answer to the question of what kind of geopolitical actor the EU is. Numerous attempts to define its role in the world (Bretherton and Vogler 2006; Manners 2010; Bachmann 2013a) are useful for structuring a respective understanding; however, they also remain selective. Whatever one makes of the EU remains entirely dependent on his/her subjective perceptions of and experiences with the EU. As such the nature of cooperation with the EU, and its position as a global actor, also remains entirely dependent on the subjective perceptions and experiences of those in charge of enacting political decisions and framework programmes, such as Cotonou or the EEAS. What sticks out in this context, are perceptions and experiences that are particularly positive and particularly negative. In this light, the inquiry about those aspects reveals much about how individuals view the EU in general and as a cooperation partner in particular.

As a general trend, it became obvious how informants perceived of a weakened geopolitical and geoeconomic role in light of the financial crisis and the growing international influence of China and other 'emerging economies'. For what is left of its international role, the categories shown in the three graphs in this chapter illustrate wider sentiments about the EU. All of them are important, and all of them are not. They are subjective perceptions. In their singularity, they may not be sufficient for defining the EU's global role. The more of such perceptions we trace, however, the more facets of the EU's global role we can identify – and the more (or less) meaning we can ascribe to these different facets. In their entirety, these subjective perceptions are decisive for shaping 'partners' interaction with the EU and thus the EU's global role. This entirety can never be fully grasped. What I offer here is a piece of the puzzle – a piece that covers some of the EU's role towards developing countries.

Conclusion
A Less Global and a More *European* Union?

The title of this book, *European External Action*, posits an engagement with the EU's active role beyond its borders. What I deliver here is of course only a small piece of a much larger puzzle of the diversity of the EU's wide and varied interactions with 'others'. As those interactions are highly heterogeneous and diversified, so is the EU as an actor itself highly heterogeneous and diversified – internally, but even more so externally. This is the case to an even greater extent regarding external perceptions of the EU. I have described the EEAS here as a heterogeneous and composite institution; the same in fact also applies to the EU as a whole. The EU covers (almost) all policy fields, yet it is not a state. It is a bureaucracy, but not just that. It also has independent agency – both internally and externally. Yet it is always a composite of different institutions and individuals with different ideological and national backgrounds. Unlike a state government, usually constituted by one party or a coalition of parties with similar positions of the political spectrum, an ideological and national balance/diversity is inscribed in the EU's structure. It is by nature heterogeneous and composite – and the building of collective agency is more complicated than in more streamlined, national, contexts. With respect to the EU's international role, the EEAS was put in place as the key means for implementing EU external relations. In addition to its own heterogeneous and composite nature, the EEAS operates in an even more heterogeneous environment, that of a two-step diplomacy, i.e. it has to find a common position with EU member states first and then enter into interactions with external partners (see Chapter 1).

Transformations from Development to Politics

Before the Lisbon Treaty, the EU's external relations were mostly restricted to areas of Commission competence; the Union's 'embassies' around the world were the Delegations of the European Commission. Diplomacy was not part of their agenda as they had no foreign political mandate from the European Council, i.e. the member states. Foreign policy remained a national affair. With the Lisbon Treaty and the EEAS, the EU is slowly emerging as a diplomatic actor on the international stage. This book traces these processes since the EEAS was formally established in 2010. Starting fieldwork in Nairobi in November 2010, just before the formal launch of the service, I accompanied the establishment of the EEAS in Kenya from the very beginning. The focus here is on the EU's political role in the diplomatic landscape of Nairobi. The EU Delegation in Nairobi is one of the largest in the world and amongst those that had been chosen to implement the changes of the Lisbon Treaty even before the official start of the EEAS. Moreover, the city is a key node for African-European relations and for the global development industry. Through the overarching presence of 'Development', Nairobi thus offers a particularly interesting setting for studying the making of a *diplomatic* role for the EU.

Over more than half a century, the EU's relations with Africa have been characterised by issues of development – and, often implied, donor-recipient hierarchies – as well as non-reciprocal trade regimes. The overarching frameworks for these relations were generalised long-term agreements (Yaoundé, Lomé, and Cotonou) with a strongly technical character about the volume and specificities of 'aid' and trade. Often denying the political nature of such 'cooperations' (Ferguson 1990), these frameworks have instilled a character of technical arrangements in Africa-European relations. Interaction was standardised and institutionalised; and European Delegations acted as implementing bureaucracies of the Commission's technical arrangements with the host governments. Such standardisation and institutionalisation of interaction have to be seen in the wider context of the EU's ambition to regulate spaces of interaction – both internally as well as externally (see Chapter 3). Frequently the EU's 'power of standardization' is a surprise to those new to the EU system or new to cooperating with the EU system (Kuus 2014, 107–8) – of course not only as regards the EU's external relations.

The push for such bureaucratisation of EU policy conduct is inscribed in its institutional evolution. In its early phase in the aftermath of World War Two, the bureaucratic modus was a more 'neutral' way of cooperating amongst the countries that had just emerged from extensive wars with each other. Even though a highly political process, EU agency sought to keep itself technical and bureaucratic so as not to raise suspicions about undermining national sovereignty and authority in political processes and fragile democracies. The Commission's role as the technical implementer of regional integration – and other collective policies – was central once the political mandate had been given by the member states. As such it also became the central negotiator and implementer of (standardised) trade agreements and development aid as regards collective relations with Africa. Its institutional mentality as a bureaucracy and a technical implementer of certain policies, and its preference for regulated spaces of interaction, has thereby also been applied in the Delegations around the world, staffed with Commission officials, and in the interaction with its external partners. The image of a technical implementer of development and trade policy thus also continues to dominate perceptions of the EU. At the same time, diplomacy and foreign policy are seen as issues within the purview of the member states (see Chapter 6). At least from the European side, the Lisbon Treaty and the EEAS were supposed to change this, transforming Commission Delegations into Union Delegations and adding a political dimension to collective external relations in an attempt to upgrade the EU's role to that of a diplomatic actor.

The enactment of such role changes, however, is a more lengthy process than the signing of the respective treaties and agreements. I employ here a critical geopolitical approach for studying the *social construction* of the making of EU diplomacy, rather than the underlying legal frameworks, in that I focus on the interactions of key individuals charged with building the EEAS and cooperating with it. As such my approach accounts both for the way European officials and institutions develop political agency and how the EU's external cooperation partners perceive the EU's transformation from that of a technical implementer to a political actor. Both processes take time – the latter more than the former. Yet also amongst European partners the anchoring of a collective dimension in the policy field of diplomacy does not occur overnight. It requires what a member state diplomat described as

> a change in mentality: we need to change the mentality of the member states to move away from doing their business on their own, but in particular we need a change in mentality on part of the EU personnel to actually become a political actor now. EU personnel have not yet caught the ball thrown at them through the Lisbon Treaty and equipping them with a lead role on political affairs of the EU. (Author interview, Nairobi, January 2012)

This transition appeared particularly difficult for those diplomats that joined the EEAS from the Commission with its institutional socialisation of bureaucratic and technical implementers. Only slowly is an *Esprit de Corps* emerging within the EEAS (Benson-Rea and Shore 2012; Juncos and Pomorska 2013b). In Kenya, its emergence, or rather the emergence of collective diplomacy, was clearly observable during this research project (2010–2014). A decisive factor was the specific situation presented by the cases of President Kenyatta and Vice-President Ruto at the International Criminal Court in The Hague, and in particular the phase after the 2013 elections.

During the fieldwork phases before the elections (2010 and 2012), the construction of a collective political role through the service appeared rather distant. The EEAS had not developed its own mark or identity – it was seen as trying to build diplomatic agency with the Commission's institutional culture of standardised and bureaucratic interaction. Such an approach is problematic in diplomacy; diplomacy is people-centred, not bureaucratic; it is dependent on the personality of individuals in charge, on tact and the sensibilities of these individuals and, of course, on personal trust and confidence. These factors are not part of the expertise in the EU's tool box of policy conduct. There is no general blueprint for diplomatic interaction; it cannot easily be standardised because it is individual and specific to each situation. In the early 1980s, in the context of initial attempts at building a collective foreign policy in the form of the European Political Cooperation (EPC), Wessels (1982, 17) pointed out that 'the common enterprise has to begin with a loose, non-binding and modest formula, through which a process of developing trust can be initiated'. Clearly, the EEAS has moved beyond being a 'loose, non-binding and modest formula', yet the importance of developing trust and sensitivity to the specific situations remains central for diplomacy. The EU's, or rather the Commission's, way of conducting policy has instead been the opposite, long prioritising the regulation of interaction spaces (see Chapter 3). Standardisation and bureaucratisation even intensified after the corruption scandal around the Santer Commission in the 1990s (see Chapter 5), leading to a further de-personalisation of policy conduct. In light of this juxtaposition between personalised diplomacy and standardised interaction, it is no surprise that the EEAS experienced a difficult start.

After the Kenyan elections in 2013, however, the EU Delegation emerged more decidedly as a political actor. The EU Ambassador had won the trust of member-state colleagues and built a personal profile, as a result of which national diplomats entrusted him with 'spearheading' the European position. He became the key European voice in the Kenyan media and a major target of criticism for the Kenyan government. In November 2012, a member state diplomat argued that, for the future of the EEAS, 'the question will be if political leaders are ready to play second role after a collective approach' (Author interview, November 2012). The situation around the elections was a major push in this direction. The EU's support for the ICC has to be seen in the context of its more general preference for the regulation of interaction spaces in terms of institutionalisation and the rule of law. It could therefore be expected that the collective European position would be strongly in favour of the ICC, and that this would create tensions with the Kenyan government's opposition to the ICC. In Kenya, the specific situation of the President and Vice-President indicted at the ICC and the personality of the EU Ambassador were thus key factors for the emergence of a collective European diplomacy. The formal institutional adjustments made several years before, i.e. the Lisbon Treaty and the EEAS, only provided the structural and institutional framework through which such collective diplomacy could potentially emerge. However, these institutional settings remained hollow structures. It was the specific local context of the ICC cases, the personality of the EU Ambassador and the personal relations and trusted interactions between diplomats that filled this structure with life and gave it meaning, and eventually also impact. This situation created the irony of collective diplomacy emerging outside, even counter to, the EU's long-established preference for regulated spaces of interaction. The first two years of building the EEAS, essentially in an attempt to extend the Commission's established, standardised-technical, way of conducting external relations to diplomacy, were of limited impact for the aspired collective political role. The major impetus for collective diplomacy came outside such regulated interaction through the special political situation in Kenya and the entrusted interactions between key personnel of the member states and the Delegation. It will remain a key challenge for the EEAS, and for the EU generally, to develop collective agency in such personalised, context-specific spaces of interaction that are beyond standardised regulation.

I want to emphasise here the value of a critical geopolitical approach and in-depth ethnographic fieldwork to the study of diplomacy. My approach has focused on the social construction of EU diplomacy in a specific site – a key site for African-

European relations and the global development industry. These relations and the EU's political agency are locally embedded. A 'localised' study of diplomatic agents and the way agency is socially constructed in particular places and through particular individuals is therefore highly appropriate. I side with Kuus' claim that the 'contingency of the empirical and the pivotal importance of individual operators must remain centre-stage in any analysis of the field' (Kuus 2014, 203). Diplomacy is *always* locally embedded as well as enacted through individuals and their relations with others. It cannot be grasped with distanced, exclusively quantitative or textual analysis. It requires ethnographic research methods and a deep immersion in the research environment to capture these factors and processes. For instance, I could only capture the importance of the place of the EU Delegation as the central location for all meetings, and consequently for centralising the role of the Delegation in perceptions of member state diplomats, through my affiliation with the Delegation. An objective of this book was to demonstrate the value of such research – and thereby to contribute not only to a better understanding of how EU diplomacy is enacted and how the EU is seen by its external cooperation partners but also more generally to the study of diplomacy as social construction.

A Critical Geopolitical Approach to the Making of EU Diplomacy in Kenya

Through such in-depth research, this book makes contributions to three larger sets of literature. *First*, it contributes to debates on the EU as a global actor and the external dimension of Europeanisation (Bretherton and Vogler 2006; Manners 2010; Bialasiewicz 2011; Wong and Hill 2012; Telò and Ponjaert 2013), focusing on the political sphere of African-European relations and the establishment of the EEAS. The distinction between *Europeanisation abroad* and *external Europeanisation* offered here is an important one. The former describes practices and processes of Europeanising the conduct of external relations between European actors. The latter describes the projection of Europeanised spaces of interaction to the relations with external 'partners'; it thus requires a more careful, critically reflexive research lens – in particular in light of Europe's imperial history. *Second*, and relatedly, it adds a detailed case study on external perceptions of the EU in Kenya to a growing research agenda inquiring into how the EU is seen from the outside (Lucarelli and Fioramonti 2010; Holland and Chaban 2014), focusing on the transition from a development to a political actor. Both debates are more at home in the disciplines of international relations and European studies than they are in geography. Yet, employing a geographic lens, this book makes a range of novel contributions to these debates – and simultaneously adds, *thirdly*, to critical geopolitics literature an engagement with the social construction of the EU's international role, with a focus on diplomacy and African-European relations. As outlined in the introduction, the critical geopolitical approach employed here studies social construction in terms of its historical and spatial conditioning, its political and economic framing, its local and international embedment, and its

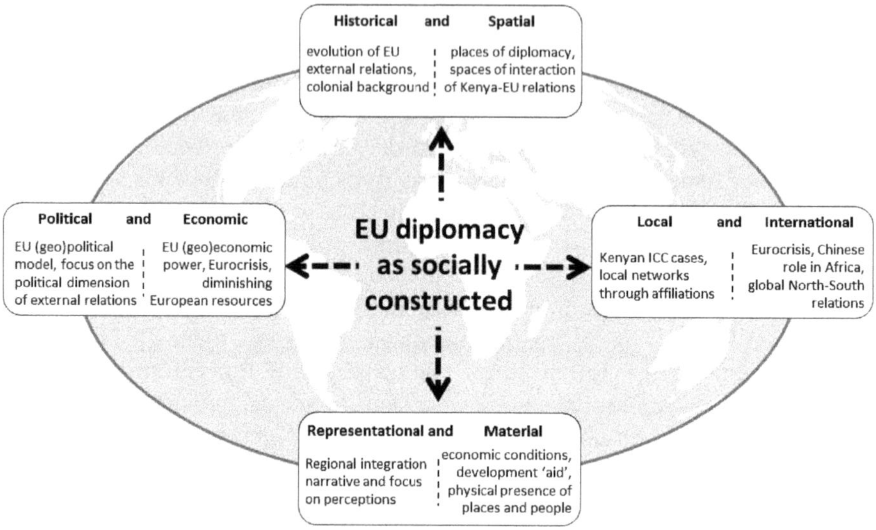

Figure C.1 A critical geopolitical approach to the making of EU diplomacy in Kenya

representationality and materiality (see Figure I.1). Figure C.1 adapts this approach to the analysis presented in this book. It provides a sketched abstraction on how each of the aspects mentioned has – at least to some extent – contributed to my analysis.

The *historical* and the *spatial* aspects are of central importance to any critical geopolitical study. The *historical* stands for the process-orientation in geographic analysis (Massey 1994). It includes the more distant historical process of European integration and external relations instruments as well as the colonial background and Kenya's post-independence evolution. However, it also includes more recent transformations, such as the processes of implementing the Lisbon Treaty and the changes in Kenya's political system during the time of the research. The *spatial* is equally important. My work conceives of space as political and of politics as spatially conditioned (Lefebvre 1991 [1974]; Dell'Agnese 2008; Belina and Michel 2011a). I introduce the notion of spaces of interaction as an overarching framework for EU policy conduct, primarily centred around networked and scalar connections and institutionalised interactions (Bachmann 2015a). Such a framework of regulated interaction spaces is in the centre of the EU's preferred modi operandi – both internally as well as externally. Yet, it quickly reaches its limits in situations outside the reach of routinisation, such as the case of personalised diplomacy presented here or the current conflicts in Syria or Ukraine. As the applicability of regulated spaces of interaction reaches its limits, so does collective EU agency. Moreover, the analysis is place-based, providing a detailed study of the sites where EU diplomacy is enacted and the social dynamics in these sites. Drawing on ethnographic research in those sites, it highlights the centrality of the place of the EU Delegation for the making of collective diplomacy as well as the importance of my physical presence in this place for capturing social dynamics between diplomats.

Beyond those two overarching, conditioning factors, I focused on the *local*, the *political* and the *representational*. Attention to the *local* is closely tied to the place-based analysis of key sites, yet it also refers to an emphasis on accounting for local contexts in the sense of a localised geopolitics (Dahlman and Ó Tuathail 2005). For this study, the local context of the Kenyan ICC cases was a decisive factor; as were my recurring fieldwork visits to Nairobi since 2005 allowing me to develop a more thorough understanding of the political climate in Kenya and the EU's role there. Also my inclusion in the (social) networks of diplomats and Kenyan foreign policy elites in Nairobi, primarily resulting from my physical presence in the Delegation and the affiliation with the University of Nairobi, played an important role for acquiring data and understanding perceptions of different informants. Analysing the EU's *political* role was a key objective of this study from the beginning. It seeks to go beyond the focus on development in African-European relations, thus studying the political dimension of EU external relations in a 'development' context. Yet, my analysis is also embedded in a wider *international* context during the time of the research project. Of particular relevance is therefore the emergence of the financial crisis in the Eurozone in 2008, the growing role of China (and others) in Africa and the wider context of global North-South relations. The focus on the *representational* is related to the ambition of capturing perceptions of those in charge of enacting Kenya-EU relations. Clearly, such perceptions are not separable from materiality or from the other factors mentioned above (local/international, political/economic, historical/spatial). All those factors together generate perceptions that, in their entirety, are decisive for filling the structures created by treaties and agreements with life and as such for shaping the practical enactment of Kenya-EU relations. In line with the critical geopolitical approach, I sought to capture such perceptions through ethnographic research and interviews in key institutions and with key individuals making and cooperating with EU diplomacy.

I connect here such critical geopolitical thought with the two sets of EU literatures mentioned above, i.e. debates on the EU as a global actor and on external perceptions of the EU. In so doing, I contribute a piece to an (endless?) puzzle trying to make sense of the EU's international identity and role. The piece of the puzzle is small but detailed. It illustrates the practices and processes of the making of EU diplomacy during the first four years of the EEAS' existence, focussing on Kenya. I use this example to also call for more such detailed case-studies in different empirical settings and point to the limitation of attempts to finitely define the EU's one and only international identity and role. The EU is a heterogeneous and composite institution; it means different things to different audiences at different time-spaces. Its identities and roles are equally heterogeneous and composite. It requires variegated engagements with these variegated identities and roles; and each engagement adds a piece to the puzzle and thus makes the whole image slightly clearer.

Regional Integration and External Perceptions

Also the results that have emerged through this research portray a variety of different ways in which the EU is seen externally. Perceptions vary depending on the time (pre Lisbon – post Lisbon, pre Kenyan elections – post Kenyan elections), on the policy field (development, political, economic), on the affiliation of the informant (African or European institution), and other factors. Some of those perceptions have been consistent over longer time periods. Others have changed. The image of the EU as a neoimperial actor, for instance, continues to persist widely amongst its African cooperation partners. Factors in the categories *Discrimination and Xenophobia* and *Paternalising Africa* are most frequently mentioned by the EU's African cooperation partners as being the most negative aspects about the EU (see Figure 8.3). Responses in these categories were related to larger macroeconomic frameworks, such as the EPA, but also to experiences of racism within Europe and increasingly restrictive border regimes and migration policies – often made responsible (directly or indirectly) for the death of African boat migrants in the Mediterranean. At the same time, however, the EU retains a substantial appeal for its civilian orientation and the historical process of regional integration on the continent (see Figure 8.2). As such the EU is also expected to function as a *Partner for Africa* and create a more equal *Global Balance*, in particular in the areas of *Good Governance, Peace and Security*, but also as regards *Trade and Economy* (see Figure 7.1 and 7.2). With regard to both the EU's role in Kenya and on a global stage, African informants' expectations of the EU in these policy fields clearly resonate with the long-established broad orientations of the civilian and normative power debates.

What remains most persistently associated with the EU, however, is the aspect of regional integration (Figure 8.1). Yet the quality of perceptions has changed. The appeal of the regional integration narrative has substantially decreased with the outbreak of the financial crisis. Regional integration is not necessarily seen any more as a desirable model to be emulated since the social and economic problems facing an integrated community, comprised of states with very different (economic) interests and capabilities, are being revealed through the crisis. Economic growth is low or negative in Europe; social tensions, inequality and xenophobia are on the rise almost everywhere on the continent. Simultaneously, other development models emerge as increasingly attractive. Most notably, and in contrast to Europe's 'normative model', this is the case with the Chinese one with its autocratic, state-orchestrated capitalism, combining a focus on large and highly visible infrastructure projects with a hands-off policy as regards conditionalities and interference in domestic affairs – not least for delivering growth rates much higher than those of a Europe in crisis.

Nevertheless, despite the loss of some of its appeal, mainly regarding the generation of economic growth, regional integration is still seen in a positive light for its commitment to peace and for the spaces of civilian interaction it has created on the European continent after long periods of war. External perceptions of the regional integration narrative are now more sober than they were before the outbreak of the crisis. Regional integration is no longer seen as a panacea, but is critically discussed in a more differentiated manner. The European model is not just seen as a possibility for learning positive lessons but also for learning negative ones about the regional integration process. As a focus group discussant remarked: 'we have to be careful to not build a European Union in East Africa but to build an East African Community' (FG1, March 2012).

Comparing these perceptions with the results of an earlier research project (before the outbreak of the crisis), it becomes obvious that the strong association of the EU with the regional integration narrative has remained, yet this association no longer entails expectations for promoting regional integration elsewhere. Figure 8.1 illustrates how, of all the informants included in this research, 91 associate the EU with regional integration but only 25 see the EU as a model. Similarly, with respect to expectations of the EU's global role, only two of the African informants mentioned the promotion of *Regional Integration* as a key role for the EU. In contrast, for European informants, this was the most frequently mentioned aspect (see Figure 7.2).

The promotion of regional integration has indeed long been a central objective of EU external relations. It is frequently called for in EU strategy papers and is one of the enduring roles for the EU in most of its geopolitical discourses. As outlined above, the appeal of the regional integration narrative, and of Europe as a model more generally, has suffered in light of the Eurocrisis and the emergence of other development models and cooperation partners, such as China. Yet with the Lisbon Treaty and the EEAS, the EU has laid the institutional foundations for playing a more active global role and for externally projecting its 'preferred world order model' (Hettne and Söderbaum 2005, 538). The EU's ambition to promote regional integration, aligned to civilian and normative power debates, I would argue, differs significantly from the previous colonial ambitions of European countries. Yet any external projection of one's own 'preferred world order model' is always contentious. Europe looks back to a long, and very problematic, history of imperialism and an equally problematic belief that it should set the global political agenda. In particular as regards African-European relations, perceptions of past and present imperialism widely persist. Whereas for European informants the difference of the EEAS' mission from the member states' colonial past is clear, for African informants it is much less obvious. The perception of European paternalism looms large. I do not want to evaluate either of these perceptions, but I do want to point to their existence and the differences. In this light, I want to emphasise the importance of taking *external* perceptions of the EU and its role in Africa seriously. Those perceptions of key individuals shaping African-European interaction are key factors for future African-European relations.

Global EUropa

Different opinions about how actively the EU should project its 'preferred world order model' externally exist not only between African and European informants. There are fundamental differences about how actively the EU (or any other polity) should promote its idealised geopolitical imaginations also amongst the European diplomats interviewed as part of this research. A member state diplomat explained how

> there are two schools of thought that collide. On the one hand, there are the Protestants in the North of Europe with their missionary ambition. They are convinced that their model of life is the best and that it is their duty to spread it around the world and to show Kenya what to do. On the other hand, there are the Catholics in the South with a practical orientation pursuing their national interests. Outside these interests, they observe what is going on and prefer to stay out. I think there are two world views facing each other here: the Protestant missionary ambition and the non-Protestant indifference. In general I think we need to ask the question if the EU should improve the world and if our conceptions are the best. (Author interview, January 2012)

More directly criticising any missionary endeavours, another member state diplomat identified a similar differentiation between the foreign political ambition of predominantly Protestant and predominantly Catholic countries in the EU. In the context of the political rivalries in Kenya and the Kenyan cases at the ICC, the diplomat argued that

> we should not come here and tell them how to cooperate with each other. It's none of our business. They prioritise their national interests and if that is what they want to do, fine. Not our cup of tea. They just want to extract European money. We should not come here and impose our point of view. We should respect the people on the other side of the table […]
>
> change here can clearly come, but it will come from internally, from the civil society. There is an active civil society and a middle class here. They are concerned with human rights and not willing to accept any more old impunity and corruption structures. But this change will come from within. Not through us, we can only support it. We are fundamentally wrong if we always believe we can change the world. It is called neo-colonialism. It's a wider problem of Protestantism in Northern Europe to missionarise. This attitude towards others in the world is problematic. (Author interview, November 2012)

The question about the level of intensity with which the EU should externally project its 'preferred world order model' is not a new one – nor is it a question restricted to the EU. It is part of a wider dilemma between interventionist and reticent foreign policy. Moreover, it is tied to the question of how the success of collective external relations, or more generally the success of the integration project, can or should be defined.

I strongly argue against measuring the success or failure of European integration in terms of the EU's global influence. It is fundamentally flawed to define the success or failure of collective diplomacy in terms of the external projection of the EU's internal spaces of interaction. European integration is first and foremost a *European* project, not a global one. Despite the historical success of creating an area of peace and relative prosperity in the post-World War Two era, European integration continues to face severe internal problems, such as high levels of unemployment, social and economic precarity, growing nationalism and xenophobia, etc. The success and failure of European integration should be measured in terms of its ability to address and resolve these pressing problems *within* the EU. From its beginning, European integration has always had significant normative components and a value-orientation towards peace, human rights, democracy, social and economic 'fairness' and non-discrimination against minorities or smaller countries. In light of this tradition, it should not, or rather it *must* not, be the objective of European integration to amass power – in particular not to amass power for the global projection of its 'preferred world order model'.

Any efforts to position the EU as a more influential global actor, for instance through the EEAS, are undermined by its internal problems and the diminishing appeal of its 'model'. International agency should not result from the EU's power to impose its model or spaces of interaction onto others; it should come from the appeal of its ideas and its own development. In the past, this appeal resulted from the EU's particular geopolitical model of regional integration, framed in civilian and normative power debates. As this research reveals, this is indeed an aspect that the EU retains appeal for, albeit to a diminishing extent. Any ambition to sustain, or even extend, some of its international influence has to be tied to this appeal, i.e. to demonstrating the advantages of the 'European model' and that people are better off in a regionally integrated union than they are without. Consequently, the key tasks for defining the EU's international role are overcoming its internal problems and overcoming its partially xenophobic and paternalistic image. An important part therefore is the cultivation of an understanding of external perceptions of the EU and its role in the world as part of a critical and reflexive engagement with the histories and complex power structures of global North-South relation. Understanding perceptions of imperialism and paternalisation, and where they come from, is a first step towards changing them.

At the same time, it is also clear that the EU cannot completely withdraw from world politics and be exclusively inward-looking. Its international connections are too extensive. The EU is still the largest integrated economy in the world and the largest donor of development assistance. Furthermore, EU member states frequently pursue similar foreign political objectives. Despite certain exceptions, often with respect to the tools to be used for achieving those objectives, the general foreign political orientation of EU member states is often similar – in particular in relation to others. As such the EU is likely to continue playing a significant international role – possibly even 'in more areas of global politics, because we often have the same interests, the same problems and the same solutions' as a member state diplomat argued (Author interview, November 2012). The often lamented (bureaucratic) inertia is thereby also a guarantor of continuity and stability of broad, long-term political orientations, often described in terms of civilian and normative power (see Chapter 2 and Bachmann and Sidaway 2009b; Manners 2010). Such stability and continuity is widely recognised and esteemed amongst the EU's international cooperation partners. An official at the East African Community, for instance, commented how 'the EU will always have a role to play, because it has very strong and stable structures. Also despite its current economic crisis, it is very stable. Measures are now taken collectively to address and fix the problem' (Author interview, November 2012).

Despite its inherent inertia and limitation of geopolitical agency, this stability and the EU's specific way of 'doing things', of how different actors in an international system interact in a civilian manner, retain a substantial appeal to others. A member state diplomat described the EU as a 'platform for discussion, interaction and articulation of different opinions and a frame to resolve different opinions without conflict or war […] and a good example on how to openly engage with problems, name them and have peaceful means for exchanging opinions' (Author interview, November 2012). In other words, it provides a space of regulated interaction for its constituents that is decidedly different from realist power politics. Similarly, in her insightful account of the production of geopolitical knowledge and authority in EU institutions, Kuus (2014, 203) argues that

> [t]he constant re-negotiation and re-imagination of compromises is the strength of EU decision-making. The long-term value of this mode of operation lies not in its specific outcomes, which are often not to anyone's liking or immediate benefit. The value of this convoluted process lies in entrenching a particular compromise-based regime of knowledge production.

This 'mode of operation', this 'convoluted process', the example of how different peoples and cultures with different opinions, interests and capabilities co-exist and cooperate without completely amalgamating, is a key factor in the EU's appeal as a geopolitical model – in particular in places that are frequently subjected to violence. Any aspect of the EU's international influence has to be rooted in this appeal and in others' perception of this appeal and demand for assistance or exchange – not in a deliberate projection of its preferred world order model. This applies especially to places where Europe has a colonial past.

Clearly, the times of an undisputed European model are over – if they ever existed. The once-guiding narrative of an 'ever-closer union' is fading in light of the financial crisis and increasing internal divisions within the EU. With it, Europe's global influence is diminishing given its evolution in demographic, political and economic terms in relation to others, in particular Asian countries. However, a more reticent global role for the EU is not a failure of European integration, nor is it a failure of collective foreign policy. Also diverging positions and different voices among EU member states and within the EEAS are not necessarily a failure. Political pluralism, to some extent even fragmentation, has always been part of the European project. Whilst it precludes streamlined unity and thus a more powerful global political role, it also characterises the EU's uniqueness and its motto of 'united in diversity'.

Bibliography

Abbott, K. W., and D. Snidal. 1998. Why States Act through Formal International Organizations. *Journal of Conflict Resolution* 42 (1):3–32.

Adebajo, A., and K. Whiteman. 2012. *The EU and Africa: From Eurafrique to Afro-Europa*. London: C Hurst & Co Publishers Ltd.

Agnew, J. 2001. How Many Europes?: The European Union, Eastward Enlargement and Uneven Development. *European Urban and Regional Studies* 8 (1):29–38.

Agnew, J. 2011. Space and Place. In *The SAGE Handbook of Geographical Knowledge*, eds. Agnew, John and David Livingston, 316–30. London: SAGE Publications Ltd.

Agnew, J., and S. Corbridge. 1995. *Mastering Space: Hegemony, Territory and International Political Economy*. London: Routledge.

Allen, D., R. Rummel, and W. Wessels. 1982. *European Political Cooperation: Towards a Foreign Policy for Western Europe*. London: Butterworth.

Allen, D., and M. Smith. 1990. Western Europe's Presence in the Contemporary International Arena. *Review of International Studies* 16 (1):19–37.

Alter, K. J. 2012. The Global Spread of European Style International Courts. *West European Politics* 35 (1):135–154.

Amin, A. 2002. Spatialities of globalisation. *Environment and Planning A* 34 (3):385–99.

Amin, A. 2004. Regions Unbound: Towards a New Politics of Place. *Geografiska Annaler: Series B, Human Geography* 86 (1):33–44.

Amnesty International. 2014. *Kenya: Victims still seeking justice for the post-election violence* Amnesty International, [cited 26/01/2015]. Available from http://www.amnesty.org/en/news/kenya-victims-still-seeking-justice-post-election-violence-2014–07–15.

Bachmann, V. 2009. Macht und/oder Modell? EU Entwicklungspolitik und differenzierte Sichtweisen Europas in Ostafrika. *Europa Regional* 17 (4):69–80.

Bachmann, V. 2010. Kenya's 'Collapse' and its Post-election Tragedy of the Commons. *Affect Journal* 2 (1):1–22.

Bachmann, V. 2011a. European Spaces of Development: Aid, Regulation and Regional Integration. In *Europe in the World: EU Geopolitics and the Transformation of European Space*, ed. Bialasiewicz, Luiza, 59–84. Farnham: Ashgate.

Bachmann, V. 2011b. Participating and Observing: Positionality and Fieldwork Relations During Kenya's Post-election Crisis. *Area* 43 (3):362–8.

Bachmann, V. 2013a. The EU's Civilian/Power Dilemma. *Comparative European Politics* 11 (4):458–80.

Bachmann, V. 2013b. The EU as a Geopolitical and Development Actor: Views from East Africa. *L'Espace Politique* 19 (1):2–22.

Bachmann, V. 2013c. International Organizations. In *The Ashgate Research Companion to Critical Geopolitics*, eds. Dodds, Klaus, Merje Kuus and Joanne Sharp, 405–20. Farnham: Ashgate.

Bachmann, V. 2013d. A Step Outside: Observations from the World's Youngest State. *Geography Compass* 7 (11):778–89.

Bachmann, V. 2015a. Global Europa, ESPON and the EU's Regulated Spaces of Interaction. *Journal of European Integration* 37 (6): 685–703.

Bachmann, V. 2015b. Spaces of Interaction: Enactments of Sociospatial Relations and an Emerging EU Diplomacy in Kenya. *Territory, Politics, Governance* (forthcoming).

Bachmann, V., and B. Belina. 2012. Crisis, Critique and the 6th International Conference of Critical Geography. *Antipode* 44 (3):555–9.

Bachmann, V., and J. D. Sidaway. 2009a. Encore – The Hour of Europe? *AAG Newsletter* 44 (6):11.

Bachmann, V., and J. D. Sidaway. 2009b. Zivilmacht Europa: A Critical Geopolitics of the European Union as a Global Power. *Transactions of the Institute of British Geographers* 34 (1):94–109.

Bachmann, V., and J. D. Sidaway. 2010. African Regional Integration and European Involvement: External agents in the East African Community *South African Geographical Journal* 92 (1):1–6.

Barnett, M. N., and M. Finnemore. 1999. The Politics, Power, and Pathologies of International Organizations. *International Organization* 53 (4):699–732.

Barnett, M. N., and M. Finnemore. 2004. *Rules for the World : International Organizations in Global Politics*. Ithaca: Cornell University Press.

Bassin, M. 1991. Russia Between Europe and Asia: The Ideological Construction of Geographical Space. *Slavic Review* 50 (1):1–17.

Bátora, J. 2013. The 'Mitrailleuse Effect': The EEAS as an Interstitial Organization and the Dynamics of Innovation in Diplomacy. *Journal of Common Market Studies* 51 (4):598–613.

Beauguitte, L., and C. Didelon-Loiseau. 2013. L'Europe vue d'ici et d'ailleurs. *L'Espace Politique* 19 (1):2–6.

Beck, U., and E. Grande. 2004. *Das kosmopolitische Europa*. Frankfurt am Main: Suhrkamp.

Belina, B., and B. Michel. 2011a. Raumproduktionen. Zu diesem Band. In *Raumproduktionen: Beiträge der Radical Geography – Eine Zwischenbilanz*, eds. Belina, Bernd and Boris Michel, 7–34. Münster: Westfälisches Dampfboot.

Belina, B., and B. Michel eds. 2011b. *Raumproduktionen: Beiträge der Radical Geography – Eine Zwischenbilanz*. Münster: Westfälisches Dampfboot.

Bendiner-Viani, G. 2013. The Big World in the Small: Layered Dynamics of Meaning-making in the Everyday. *Environment and Planning D: Society and Space* 31 (4):708–26.

Benson-Rea, M., and C. Shore. 2012. Representing Europe: The Emerging 'Culture' of EU Diplomacy. *Public Administration* 90 (2):480–96.
Berger, C. 2012. *The European External Action Service: Increasing Coherence in the Crisis Management of the European Union?* Marburg: Tectum.
Bialasiewicz, L. ed. 2011. *Europe in the World: EU Geopolitics and the Transformation of European Space*. Aldershot: Ashgate.
Bialasiewicz, L., C. Dahlman, G. M. Apuzzo, F. Ciuta, A. Jones, C. Rumford, R. Wodak, J. Anderson, and A. Ingram. 2009. Interventions in the New Political Geographies of the European 'Neighborhood'. *Political Geography* 28 (2):79–89.
Biscop, S. and R. Coolsaet. 2003. The World is the Stage – A Global Security Strategy for the European Union'. *Policy Paper*. Notre Europe. Paris. 8.
Blumer, H. 1954. What Is Wrong with Social Theory. *American Sociological Review* 18:3–10.
Boedeltje, F., and H. v. Houtum. 2011. Brussels is Speaking: The Adverse Speech Geo-Politics of the European Union Towards its Neighbours. *Geopolitics* 16 (1):130–45.
Börzel, T. A., and T. Risse. 2012a. From Europeanisation to Diffusion: Introduction. *West European Politics* 35 (1):1–19.
Börzel, T. A., and T. Risse. 2012b. When Europeanisation Meets Diffusion: Exploring New Territory. *West European Politics* 35 (1):192–207.
Bowen, G. 2006. Grounded Theory and Sensitizing Concepts. *International Journal of Qualitative Methods* 5 (3):12–23.
Bradbury, H., and B. M. B. Lichtenstein. 2000. Relationality in Organizational Research: Exploring The Space Between. *Organization Science* 11 (5):551–64.
Brenner, N., and S. Elden. 2009. Henri Lefebvre on State, Space, Territory. *International Political Sociology* 3 (4):353–77.
Bretherton, C., and J. Vogler. 2006. *The European Union as a Global Actor*. London: Routledge.
Briet, L.2015. *Welcome*. Delegation of the European Union to Kenya, [cited 16/01/2015]. Available from http://eeas.europa.eu/delegations/kenya/about_us/welcome/index_en.htm.
Bull, H. 1982. Civilian Power Europe: A Contradiction in Terms? *Journal of Common Market Studies* 21 (1/2):149–64.
Callon, M., and B. Latour. 1981. Unscrewing the Big Leviathan: How Actors Macro-structure Reality and How Sociologists Help Them to Do o. In *Advances in Social Theory and Methodology: Toward an Integration of Micro- and Macro-Sociologies*, eds. Knorr-Cetina, K. and Aaron Victor Cicourel, 277–303. Boston: Routledge & Kegan Paul.
Cameron, F. 2011. The EU's External Action Service – Golden or Missed Opportunity? In *The High Representative for the EU Foreign and Security Policy – Review and Prospects*, eds. Müller-Brandeck-Bocquet, Gisela and Carolin Rüger, 235–58. Baden-Baden: Nomos.

Campbell, H. 2008. China in Africa: Challenging US Global Hegemony. *Third World Quarterly* 29 (1):89–105.
Campbell, L. M., N. J. Gray, Z. A. Meletis, J. G. Abbott, and J. J. Silver. 2006. Gatekeepers and Keymasters: Dynamic Relationships of Access in Geographical Fieldwork. *Geographical Review* 96 (1):97–121.
Carbone, M. 2007. *The European Union and International Development: The Politics of Foreign Aid*. London: Routledge.
Carta, C. 2012. *The European Union Diplomatic Service : Ideas, Preferences and Identities*. London: Routledge.
Cebeci, M. 2012. European Foreign Policy Research Reconsidered: Constructing an 'Ideal Power Europe' through Theory? *Millennium – Journal of International Studies* 40 (3):563–83.
Chaban, N., O. Elgström, S. Kelly, and L. S. Yi. 2013. Images of the EU beyond its Borders: Issue-Specific and Regional Perceptions of European Union Power and Leadership. *Journal of Common Market Studies* 51 (3):433–51.
Chang, H.-j. 2004. *Reclaiming Development: An Alternative Economic Policy Manual*. London: Zed Books.
Chazan, N., P. Lewis, R. Mortimer, D. Rothchild, and S. Stedman. 1999. *Politics and Society in Contemporary Africa*. 3rd ed. Boulder: Lynne Rienner.
Cheneval, F., and F. Schimmelfennig. 2013. The Case for Demoicracy in the European Union*. *Journal of Common Market Studies* 51 (2):334–50.
Ciuta, F., and I. Klinke. 2010. Lost in Conceptualization: Reading the 'New Cold War' with critical geopolitics. *Political Geography* 29 (6):323–32.
Clark, J., and A. Jones. 2008. The Spatialities of Europeanisation: Territory, Government and Power in 'Europe'. *Transactions of the Institute of British Geographers* 33 (3):300–318.
Clark, J., and A. Jones. 2009. Europeanisation and Its Discontents. *Space and Polity* 13 (3):193–212.
Conway, M., and K. K. Patel. 2010. *Europeanization in the Twentieth century: Historical Approaches*. Basingstoke: Palgrave Macmillan.
Coudenhove-Kalergi, R. 1931. Informationsschrift über die Paneuropa-Bewegung. Wien.
Coudenhove-Kalergi, R. 1934. *Europa Erwachet!* Wien: Paneuropa-Verlag.
Cresswell, T. 2004. *Place: A Short Introduction*. Oxford: Blackwell.
Dahlman, C., and G. Ó Tuathail. 2005. Broken Bosnia: The Localized Geopolitics of Displacement and Return in Two Bosnian Places. *Annals of the Association of American Geographers* 95 (3):644–62.
Davies, C. A. 2008. *Reflexive Ethnography: A Guide to Researching Selves and Others*. 2. ed. London: Routledge.
Davies, N. 1997. *Europe: A History*. London: Pimlico.
Degnen, C. 2013. 'Knowing', Absence, and Presence: The Spatial and Temporal Depth of Relations. *Environment and Planning D: Society and Space* 31 (3):554–70.

Delanty, G., and C. Rumford. 2005. *Rethinking Europe: Social Theory and the Implications of Europeanization*. London: Routledge.

Deleuze, G. 1995. *Negotiations: 1972–1990*. New York: Columbia University Press.

Dell'Agnese, E. 2008. Geo-graphing: Writing Worlds. In *The SAGE handbook of political geography*, eds. Cox, Kevin R., Murray Low and Jennifer Robinson, 439–53. Los Angeles: SAGE.

Dialer, D. 2007. *Die EU-Entwicklungspolitik im Brennpunkt. Eine Analyse der politischen Dimension des Cotonou-Abkommens*. Frankfurt: Brandes & Apsel.

Diez, T. 2005. Constructing the Self and Changing Others: Reconsidering 'Normative Power Europe'. *Millennium – Journal of International Studies* 33 (3):613–36.

Dittmer, J., and K. Dodds. 2008. Popular Geopolitics Past and Future: Fandom, Identities and Audiences. *Geopolitics* 13 (3):437–57.

Dodds, K., M. Kuus, and J. Sharp. 2013. *The Ashgate Research Companion to Critical Geopolitics*. Farnham: Ashgate.

Doyle, M. 2008. Kenya's Geographic and Political Rift. *BBC News*, 28/01/2008.

Duchêne, F. 1972. Europe's Role in World Peace. In *Europe Tomorrow. Sixteen Europeans Look Ahead*, ed. Mayne, Richard, 32–47. London: Fontana/Collins.

Duchêne, F. 1973. The European Community and the Uncertainties of Interdependence In *A Nation Writ Large? Foreign Policy Problems before the European Community*, eds. Kohnstamm, Max and Wolfgang Hager, 1–21. Basingstoke: Macmillian.

Duke, S. 2009. Providing for European-Level Diplomacy after Lisbon: The Case of the European External Action Service. *The Hague Journal of Diplomacy* 4 (2):211–33.

EC. 2007a. Republic of Kenya – European Community. Country Strategy Paper and Indicative Programme for the period 2008–2013: European Commission.

——. 2007b. Treaty of Lisbon. *Official Journal of the European Union* C 306 (17 December 2007).

——. 2013a. *Economic Partnerships*. European Commission [cited 18/12/2013]. Available from http://ec.europa.eu/trade/policy/countries-and-regions/development/economic-partnerships/index_en.htm.

——. 2013b. *EU Budget and External Cooperation*. European Commission [cited 18/12/2013]. Available from http://ec.europa.eu/europeaid/how/finance/mff/eu-budget_en.htm.

——. 2014a. *Development Policies*. European Commission [cited 21/07/2014]. Available from http://ec.europa.eu/europeaid/what/development-policies/index_en.htm.

——. 2014b. *European Development Fund (EDF)*. European Commission [cited 04/08/2014]. Available from http://ec.europa.eu/europeaid/how/finance/edf_en.htm.

EEAS. 2013a. *Common Foreign and Security Policy (CFSP) of the European Union*. European External Action Service, [cited 18/12/2013]. Available from http://eeas.europa.eu/cfsp/index_en.htm.

——. 2013b. *Ongoing Missions and Operations*. European External Action Service, [cited 18/12/2013]. Available from http://eeas.europa.eu/csdp/missions-and-operations/index_en.htm.

——. 2014a. *The Common Foreign and Security Policy (CFSP)*. European External Action Service, [cited 22/07/2014]. Available from http://eeas.europa.eu/delegations/council_europe/key_eu_policies/common_foreign_security_policy/index_en.htm.

——. 2014b. *Kenya & the EU – Partners For Life*. European External Action Service, [cited 12/03/2014]. Available from http://eeas.europa.eu/delegations/kenya/eu_kenya/index_en.htm.

——. 2014c. *Kenya & the EU – Trade*. European External Action Service, [cited 12/03/2014]. Available from http://eeas.europa.eu/delegations/kenya/eu_kenya/trade_relation/index_en.htm.

——. 2014d. *Remarks by EU High Representative Catherine Ashton at the European Parliament in the debate on foreign and defence policy*. European External Action Service, [cited 10/08/2014]. Available from http://eeas.europa.eu/statements/docs/2014/140403_03_en.pdf.

——. 2015a. *EU Relations with North Africa, Middle East and Arabian Peninsula*. European External Action Service, [cited 24/07/2015]. Available from http://www.eeas.europa.eu/mideast/index_en.htm.

——. 2015b. *Staff at EU Delegations*. European External Action Service (Information obtained upon request from the EEAS' Office of Human Resource on 23/07/2015).

Elden, S. 2005. Missing the Point: Globalization, Deterritorialization and the Space of the World. *Transactions of the Institute of British Geographers* 30 (1):8–19.

Elden, S. 2007. Governmentality, Calculation, Territory. *Environment and Planning D: Society and Space* 25 (3):562–80.

Elden, S. 2013. Secure the Volume: Vertical Geopolitics and the Depth of Power. *Political Geography* 34:35–51.

Enengel, B., A. Muhar, M. Penker, B. Freyer, S. Drlik, and F. Ritter. 2012. Co-production of Knowledge in Transdisciplinary Doctoral Theses on Landscape Development – An Analysis of Actor Roles and Knowledge Types in Different Research Phases. *Landscape and Urban Planning* 105 (1–2):106–17.

England, K. 1994. Getting Personal: Reflexivity, Positionality, and Feminist Research. *Professional Geographer* 46 (1):80–89.

Eriksen, E. O. 2006. The EU – A Cosmopolitan Polity? *Journal of European Public Policy* 13 (2):252–69.

Escobar, A. 1992. Reflections on 'Development': Grassroot approaches and Alternative Politics in the Third World. *Futures* 24:411–36.

EU.2013. *The European Report on Development (ERD) 2013 – Post 2015: Global Action for an Inclusive and Sustainable Future*. European Union, [cited 21/07/2014]. Available from http://www.erd-report.eu/erd/report_2012/index.html.

Evans, M. 1988. Participant Observation – The Researcher as a Research Tool. In *Qualitative Methods in Human Geography*, ed. Smith, John Eyles and David, 197–218. Cambridge: Polity Press.

Exadaktylos, T., and C. M. Radaelli. 2009. Research Design in European Studies: The Case of Europeanization*. *Journal of Common Market Studies* 47 (3):507–30.

Eyles, J. 1988. Interpreting the Geographical World. Qualitative Approaches in Geographical Research. In *Qualitative Methods in Human Geography*, ed. Smith, John Eyles and David, 1–16. Cambridge: Polity Press.

Falkner, G. 2013. The JCMS Annual Review Lecture: Is the European Union Losing Its Credibility? *JCMS: Journal of Common Market Studies* 51:13–30.

Fanon, F. 1961. *Les damnés de la terre*. Paris: Maspero.

Ferguson, J. 1990. *The Anti-politics Machine: Development, Depoliticization, and Bureaucratic Power in Lesotho*. Cambridge: Cambridge University Press.

Fioramonti, L., and A. Poletti. 2008. Facing the Giant: Southern Perspectives on the European Union. *Third World Quarterly* 29 (1):167–80.

Flockhart, T. 2010. Europeanization or EU-ization? The Transfer of European Norms across Time and Space. *Journal of Common Market Studies* 48 (4):787–810.

Forsberg, T. 2011. Normative Power Europe, Once Again: A Conceptual Analysis of an Ideal Type. *Journal of Common Market Studies* 49 (6):1183–1204.

Friedrichs, J., and F. Kratochwil. 2009. On Acting and Knowing: How Pragmatism Can Advance International Relations Research and Methodology. *International Organization* 63 (04):701–31.

Galtung, J. 1973. *The European Community: A Superpower in the Making*. Oslo: Universitetforlaget.

Gans, H. J. 1982. The Participant Observer as a Human Being: Observations on the Personal Aspects of Fieldwork. In *Field Research: A Sourcebook and Field Manual*, ed. Burgess, Robert G., 53–61. London: Routledge.

Gehring, T., S. Oberthür, and M. Mühleck. 2013. European Union Actorness in International Institutions: Why the EU is Recognized as an Actor in Some International Institutions, but Not in Others. *JCMS: Journal of Common Market Studies* 51 (5):849–65.

Germond, B. 2013. The European Union at the Horn of Africa: The Contribution of Critical Geopolitics to Piracy Studies. *Global Policy* 4 (1):80–85.

Geuijen, K., P. t. Hart, S. Princen, and K. Yesilkagit. 2008. *The New Eurocrats: National Civil Servants in EU Policy-making*. Amsterdam: Amsterdam University Press.

Gibb, R. 2006. The European Union's 'Everything But Arms' Development Initiative and Sugar: Preferential Access or Continued Protectionism? *Applied Geography* 26 (1):1–17.

Grataloup, C. 2009. *L'invention des continents*. Paris: Larousse.

Graziano, P., and M. P. Vink. 2008. *Europeanization: New Research Agendas*. Basingstoke: Palgrave Macmillan.

Gros, D.2014. *The Government Europe Deserves?* Centre for European Policy Studies: Commentary 17 September 2014, [cited 12/02/2015]. Available from http://www.ceps.eu/book/government-europe-deserves.

Hancock, G. 1989. *Lords of Poverty: The Power, Prestige, and Corruption of the International Aid Business*. New York: Atlantic Monthly Press.

Harvey, D. 1969. *Explanation in Geography*. London: Arnold.

Harvey, D. 1973. *Social Justice and the City*. London: Arnold.

Harvey, D. 1989. *The Condition of Postmodernity: An Enquiry into the Origins of Cultural Change*. Oxford: Blackwell.

Harvey, D. 1996. *Justice, Nature and the Geography of Difference*. Cambridge: Blackwell.

Harvey, D. 2006. Space as a Keyword. In *David Harvey*, eds. Castree, Noel and Derek Gregory, 270–93. Malden: Blackwell.

Hemming, P. J. 2008. Mixing Qualitative Research Methods in Children's Geographies. *Area* 40 (2):152–62.

Henderson, K. 1999. *Back to Europe: Central and Eastern Europe and the European Union*. London; Philadelphia: UCL Press.

Herod, A. 1999. Reflections on Interviewing Foreign Elites: Praxis, Positionality, Validity, and the Cult of the Insider. *Geoforum* 30 (4):313–27.

Hettne, B., and F. Söderbaum. 2005. Civilian Power or Soft Imperialism? The EU as a Global Actor and the Role of Interregionalism. *European Foreign Affairs Review* 10 (4):535–52.

Hill, C. 1993. The Capability-expectations Gap, or Conceptualizing Europe's International Role. *Journal of Common Market Studies* 31 (3):305–28.

Hill, C., and W. Wallace. 1996. Introduction: Actors and Actions. In *The actors in Europe's foreign policy*, ed. Hill, Christopher, 1–18. London: Routledge.

Holden, P. 2009. *In Search of Structural Power: EU Aid Policy as a Global Political Instrument*. Aldershot: Ashgate.

Holden, P. 2012. The End of Solidarité? The Euro Crisis and the European Union Model of Integration. In *Regions and Crisis: New Challenges for Contemporary Regionalism*, ed. Fioramonti, Lorenzo, 67–84.

Holland, M., and N. Chaban eds. 2014. *Communicating Europe in Times of Crisis: External Perceptions of the European Union*. Basinkstoke: Palgrave Macmillan.

Hooper, B., and O. Kramsch. 2007. Post-Colonising Europe: The Geopolitics of Globalisation, Empire and Borders: Here and There, Now and Then. *Tijdschrift voor Economische en Sociale Geografie* 98 (4):526–34.

Houtum, H. v., and F. Boedeltje. 2009. Europe's Shame: Death at the Borders of the EU. *Antipode* 41 (2):226–30.

Howorth, J. 2010. The EU as a Global Actor: Grand Strategy for a Global Grand Bargain? *Journal of Common Market Studies* 48 (3):455–74.

Howorth, J. 2011. The 'New Faces' of Lisbon: Assessing the Performance of Catherine Ashton and Herman van Rompuy on the Global Stage. *European Foreign Affairs Review* 16 (3):303–23.

Hugill, P. J. 2008. German Great-power Relations in the Pages of Simplicissimus, 1896–1914. *The Geographical Review* 98 (1):1–23.
Hugill, P. J. 2009. The American Challenge to British Hegemony, 1861–1947. *Geographical Review* 99 (3):403–25.
Hugill, P. J., and V. Bachmann. 2005. The Route to the Techno-Industrial World Economy and the Transfer of German Organic Chemistry to America Before, During, and Immediately After World War I. *Comparative Technology Transfer and Society* 3 (2):158–86.
Hulse, M. 2014. Actorness Beyond the European Union: Comparing the International Trade Actorness of SADC and ECOWAS. *JCMS: Journal of Common Market Studies* 52 (3):547–65.
Hurt, S. R. 2003. Co-operation and Coercion? The Cotonou Agreement Between the European Union and ACP States and the End of the Lomé Convention. *Third World Quarterly* 24 (1):161–76.
Hurt, S. R. 2012. The EU-SADC Economic Partnership Agreement Negotiations: 'Locking in' the Neoliberal Development Model in Southern Africa? *Third World Quarterly* 33 (3):495–510.
Hyde-Price, A. 2006. 'Normative' Power Europe: A Realist Critique. *Journal of European Public Policy* 13 (2):217–34.
ICG.2012. *Kenya: Impact of the ICC Proceedings* International Crisis Group, [cited 26/01/2015]. Available from http://www.crisisgroup.org/~/media/Files/africa/horn-of-africa/kenya/B084%20Kenya%20----%20Impact%20of%20the%20ICC%20Proceedings.pdf.
Jackson, J. T. 2005. *The Globalizers: Development Workers in Action*. Baltimore: Johns Hopkins University Press.
Jacoby, W., and S. Meunier. 2010. Europe and the Management of Globalization. *Journal of European Public Policy* 17 (3):299–317.
Jensen, O. B., and T. Richardson. 2004. *Making European Space: Mobility, Power and Territorial Identity*. London; New York: Routledge.
Jessop, B., N. Brenner, and M. Jones. 2008. Theorizing Sociospatial Relations. *Environment and Planning D: Society and Space* 26 (3):389–401.
Jetschke, A., and P. Murray. 2012. Diffusing Regional Integration: The EU and Southeast Asia. *West European Politics* 35 (1):174–91.
Jick, T. D. 1979. Mixing Qualitative and Quantitative Methods: Triangulation in Action. *Administrative Science Quarterly* 24 (4):602–11.
Jones, A. 2006. Narrative-Based Production of State Spaces for International Region Building: Europeanization and the Mediterranean. *Annals of the Association of American Geographers* 96 (2):415–31.
Jones, A., and J. Clark. 2008. Europeanisation and Discourse Building: The European Commission, European Narratives and European Neighbourhood Policy. *Geopolitics* 13 (3):545–71.
Jones, A., and J. Clark. 2010. *The Spatialities of Europeanization: Power, Governance and Territory in Europe*. London: Routledge.

Jones, E. 2012. European Crisis, European Solidarity. *Journal of Common Market Studies* 50:53–67.
Jones, L., and D. Sage. 2010. New Directions in Critical Geopolitics: An Introduction. *GeoJournal* 75 (4):315–25.
Jones, M. 2009. Phase Space: Geography, Relational Thinking, and Beyond. *Progress in Human Geography* 33 (4):487–506.
Jones, M., and B. Jessop. 2010. Thinking State/Space Incompossibly. *Antipode* 42 (5):1119–49.
Jorgensen, D. L. 1989. *Participant Observation: A Methodology for Human Studies*. 2nd printing . ed. Newbury Park, Calif. ; London: Sage.
Juncos, A. E. 2011. Europeanization by Decree? The Case of Police Reform in Bosnia. *JCMS: Journal of Common Market Studies* 49 (2):367–89.
Juncos, A. E., and K. Pomorska. 2013a. 'In the Face of Adversity': Explaining the attitudes of EEAS officials *vis-à-vis* the new service. *Journal of European Public Policy* 20 (9):1332–49.
Juncos, A. E., and K. Pomorska. 2013b. Manufacturing Esprit de Corps: The Case of the European External Action Service. *JCMS: Journal of Common Market Studies* 52 (2):302–19.
Kagan, R. 2003. *Of Paradise and Power: America and Europe in the New World Order*. 1st ed. New York: Alfred A. Knopf: Distributed by Random House.
Kaldor, M. 2012. The EU as a New Form of Political Authority: The Example of the Common Security and Defence Policy. *Global Policy* 3:79–86.
Kearns, R. 2000. Being There: Research through Observing and Participating. In *Qualitative Research Methods in Human Geography*, ed. Hay, Iain, 103–21. Oxford: Oxford University Press.
Kelley, J. G. 2004. *Ethnic Politics in Europe: The Power of Norms and Incentives*. Princeton, N.J. ; Oxford: Princeton University Press.
Kennedy, L. 2006. Spectres of Comparison American Studies and the United States of the West. *Comparative American Studies* 4 (2):135–50.
Keukeleire, S., and B. Hooijmaaijers. 2014. The BRICS and Other Emerging Power Alliances and Multilateral Organizations in the Asia-Pacific and the Global South: Challenges for the European Union and Its View on Multilateralism. *JCMS: Journal of Common Market Studies* 52 (3):582–99.
Kiamba, A., and V. Bachmann. 2015. Kenya-EU relations: Perceptions and Expectations. In *Perceptions of the EU in Eastern Europe and Sub-Saharan Africa: Looking In from the Outside*, eds. Bachmann, Veit and Martin Müller. Basingstoke: Palgrave.
Kipfer, S., P. Saberi, and T. Wieditz. 2013. Henri Lefebvre: Debates and Controversies. *Progress in Human Geography* 37 (1):115–34.
Kirste, K., and H. Maull. 1996. Zivilmacht und Rollentheorie. *Zeitschrift für Internationale Beziehungen* 3 (2):283–312.
Kuus, M. 2007. *Geopolitics Reframed: Security and Identity in Europe's Eastern Enlargement*. New York: Palgrave Macmillan.

Kuus, M. 2010a. Critical Geopolitics. In *The International Studies Encyclopedia. Blackwell*, Volume II, ed. Denemark, Robert, 683–701.
Kuus, M. 2010b. EUrope and the Baroque. *Environment and Planning D: Society and Space* 28 (3):381–87.
Kuus, M. 2011a. Bureaucracy and Place: Expertise in the European Quarter. *Global Networks: A Journal of Transnational Affairs* 11 (4):1–19.
Kuus, M. 2011b. Policy and Geopolitics: Bounding Europe in EUrope. *Annals of the Association of American Geographers* 101 (5):1140–55.
Kuus, M. 2011c. Whose regional expertise? Political Geographies of Knowledge in the European Union. *European Urban and Regional Studies* 18 (3):275–88.
Kuus, M. 2012. Foreign Policy and Ethnography: A Sceptical Intervention. *Geopolitics*:1–17.
Kuus, M. 2014. *Geopolitics and Expertise: Knowledge and Authority in European Diplomacy*. Chichester: Wiley.
Laclau, E., and C. Mouffe. 1985. *Hegemony and Socialist Strategy: Towards a Radical Democratic Politics*. London: Verso.
Langbein, J., and K. Wolczuk. 2012. Convergence Without membership? The Impact of the European Union in the Neighbourhood: Evidence from Ukraine. *Journal of European Public Policy* 19 (6):863–81.
Larsen, H. 2014. The EU as a Normative Power and the Research on External Perceptions: The Missing Link. *JCMS: Journal of Common Market Studies* 52(4): 896–910.
Law, J. 2004. And If the Global Were Small and Noncoherent? Method, Complexity, and the Baroque. *Environment and Planning D: Society and Space* 22 (1):13–26.
Lefebvre, H. 1970. *La révolution urbaine*. (Paris): Gallimard.
Lefebvre, H. 1973. *La survie du capitalisme: la re-production des rapports de production*. Paris: Anthropos.
Lefebvre, H. 1974. *La production de l'espace*. Paris: Ed. Anthropos.
Lefebvre, H. 1991 [1974]. *The Production of Space*. Oxford: Blackwell.
Lefebvre, H., N. Brenner, and S. Elden. 2009. *State, Space, World: Selected Essays*. Minneapolis: Univ. of Minnesota Press.
Leftwich, A. 1995. Bringing Politics Back In: Towards a Model of the Developmental State. *The Journal of Development Studies* 31 (3):400–427.
Lenz, T. 2012. Spurred Emulation: The EU and Regional Integration in Mercosur and SADC. *West European Politics* 35 (1):155–73.
Lincoln, J. R. 1982. Intra- (and Inter-) Organizational Networks. In *Research in the Sociology of Organizations*, ed. Bacharach, Samuel B., 1–38. Greenwich: Jai Press.
Lodge, J. 1993. From Civilian Power to Speaking with a Common Voice: The Transition to a CFSP. In *The European Community and the Challenge of the Future*, ed. Lodge, Juliet. London: Pinter Publishers.
Lucarelli, S., and L. Fioramonti. 2010. *External Perceptions of the European Union as a Global Actor*. London: Routledge.

Maina, B., and A. Muliro. 2008. Post-election Crisis: Kenya Can Learn From Flying Geese. *The Daily Nation*, 01/02/2008:18–19.

Mamadouh, V. 2010. Critical Geopolitics at a (not So) Critical Junction. In Jones, Laura and Daniel Sage: New Directions in Critical Geopolitics: An Introduction. *GeoJournal* 75 (4):315–25.

Mandel, J. L. 2003. Negotiating Expectations in the Field: Gatekeepers, Research Fatigue and Cultural Biases. *Singapore Journal of Tropical Geography* 24 (2):198–210.

Manners, I. 2002. Normative Power Europe: A Contradiction in Terms? *Journal of Common Market Studies* 40 (2):235–58.

Manners, I. 2008. The Normative Ethics of the European Union. *International Affairs* 84 (1):45–60.

Manners, I. 2010. Global Europa: Mythology of the European Union in World Politics. *Journal of Common Market Studies* 48 (1):67–87.

Manners, I. 2013. European Communion: Political Theory of European Union. *Journal of European Public Policy* 20 (4):473–94.

Marston, S. A., J. P. Jones, and K. Woodward. 2005. Human Geography Without Scale. *Transactions of the Institute of British Geographers* 30 (4):416–32.

Massey, D. 1994. *Space, Place, and Gender*. Minneapolis: University of Minnesota Press.

Massey, D. 1999. Space-Time, 'Science' and the Relationship between Physical Geography and Human Geography. *Transactions of the Institute of British Geographers* 24 (3):261–76.

Massey, D. 2004a. Geographies of Responsibility. *Geografiska Annaler: Series B, Human Geography* 86 (1):5–18.

Massey, D. 2004b. The Political Challenge of Relational Space: Introduction to the Vega Symposium. *Geografiska Annaler: Series B, Human Geography* 86 (1):3–3.

Mathiason, J. 2007. *Invisible Governance: International Secretariats in Global Politics*. Bloomfield: Kumarian Press.

Maull, H. W. 1993. Civilian Power: The Concept and Its Relevance for Security Issues. In *Mapping the Unknown, Towards a New World Order*, eds. Babic, Lidija and Bo Huldt, 115–31. Stockholm: Swedish Institute of International Affairs.

Maull, H. W. 2005. Europe and the New Balance of Global Order. *International Affairs* 81 (4):775–99.

McCann, E. 2010. Urban Policy Mobilities and Global Circuits of Knowledge: Toward a Research Agenda. *Annals of the Association of American Geographers* 101 (1):107–30.

McConnell, F., T. Moreau, and J. Dittmer. 2012. Mimicking State Diplomacy: The Legitimizing Strategies of Unofficial Diplomacies. *Geoforum* 43 (4):804–14.

Megoran, N. 2006. For Ethnography in Political Geography: Experiencing and Re-imagining Ferghana Valley Boundary Closures. *Political Geography* 25 (6):622–40.

Meinig, D. 2004. *The Shaping of America – A Geographical Perspective on 500 Years of History: Volume 4 Global America 1915–2000*. New Haven: Yale University Press.

Men, J. 2011. *China and the European Union in Africa: Partners or Competitors?* Farnham: Ashgate.

Menon, A. 2009. Empowering Paradise? The ESDP at Ten. *International Affairs* 85 (2):227–46.

Menon, A. 2014. The JCMS Annual Review Lecture Divided and Declining? Europe in a Changing World. *JCMS: Journal of Common Market Studies* 52:5–24.

Michalski, A. 2013. Europeanization of National Foreign Policy: The Case of Denmark's and Sweden's Relations with China. *JCMS: Journal of Common Market Studies* 51 (5):884–900.

Moïsi, D. 2009. *The Geopolitics of Emotion*. New York: Doubleday.

Moisio, S. 2011. Geographies of Europeanization: The EU's Spatial Planning as a Politics of Scale. In *Europe in the World: EU Geopolitics and the Transformation of European Space*, ed. Bialasiewicz, Luiza, 19–39. Aldershot: Ashgate.

Moisio, S. 2015. Geopolitics/Critical Geopolitics. In *The Wiley-Blackwell Companion to Political Geography*, eds. Agnew, John, Virginie Mamadouh, Anna Secor and Joanne Sharp. Oxford: Wiley-Blackwell.

Moisio, S., V. Bachmann, L. Bialasiewicz, E. dell'Agnese, J. Dittmer, and V. Mamadouh. 2013. Mapping the Political Geographies of Europeanization: National Discourses, External Perceptions and the Question of Popular Culture. *Progress in Human Geography* 37 (6):737–61.

Moisio, S., and J. Luukkonen. 2014. European Spatial Planning as Governmentality: An Inquiry into Rationalities, Techniques, and Manifestations. *Environment and Planning C: Government and Policy*.

Moravcsik, A., and M. A. Vachudova. 2003. National Interests, State Power, and EU Enlargement. *East European Politics and Societies* 17 (1):42–57.

Moumoutzis, K. 2011. Still Fashionable Yet Useless? Addressing Problems with Research on the Europeanization of Foreign Policy. *Journal of Common Market Studies* 49 (3):607–29.

Mügge, D. 2014. Europe's Regulatory Role in Post-crisis Global Finance. *Journal of European Public Policy* 21 (3):316–26.

Müller, M. 2008. Reconsidering the Concept of Discourse for the Field of Critical Geopolitics: Towards Discourse as Language and Practice. *Political Geography* 27 (3):322–38.

Müller, M. 2010. Doing Discourse Analysis in Critical Geopolitics. *L'Espace Politique* 12 (3):1–18.

Müller, M. 2012. Opening the Black Box of the Organization: Socio-material Practices of Geopolitical Ordering. *Political Geography* 31 (6):379–88.

Murphy, A. B. 2013. Trapped in the Logic of the Modern State System? European Integration in the Wake of the Financial Crisis. *Geopolitics* 18 (3):705–23.

Murray, P., and E. Moxon-Browne. 2013. The European Union as a Template for Regional Integration? The Case of ASEAN and Its Committee of Permanent Representatives. *JCMS: Journal of Common Market Studies* 51 (3):522–37.

Myers, G. 2015. A World-Class City-Region? Envisioning the Nairobi of 2030. *American Behavioral Scientist* (forthcoming).

Neumann, I. B. 2012. *At Home with the Diplomats: Inside a European Foreign Ministry*. Ithaca: Cornell University Press.

Nicholls, W. 2009. Place, Networks, Space: Theorising the Geographies of Social Movements. *Transactions of the Institute of British Geographers* 34 (1):78–93.

Nicolaïdis, K. 2013. European Demoicracy and Its Crisis. *Journal of Common Market Studies* 51 (2):351–69.

Nicolaïdis, K., and R. Howse. 2002. 'This is my EUtopia … ': Narrative as Power. *Journal of Common Market Studies* 40 (4):767–92.

Nixson, F. 2007. Aid, Trade and Economic Development: The EU and the Developing World. In *The Economics of the European Union: Policy and Analysis*, eds. Artis, Mike and Frederick Nixson, 322–53. Oxford; New York: Oxford University Press.

Nohria, N. 1992. Is a Network Perspective a Useful Way of Studying Organizations? In *Networks and Organizations: Structure, Form, and Action*, eds. Nohria, Nitin and Robert G Eccles, 1–22. Boston, Mass.: Harvard Univ. Press.

O'Loughlin, J., G. Ó Tuathail, and V. Kolossov. 2005. Russian Geopolitical Culture and Public Opinion: The Masks of Proteus Revisited. *Transactions of the Institute of British Geographers* 30 (3):322–35.

Ó Tuathail, G. 1996. *Critical Geopolitics*. Minneapolis: University of Minnesota Press.

Ó Tuathail, G. 1999. Understanding Critical Geopolitics: Geopolitics and Risk Society. *Journal of Strategic Studies* 22 (2–3):107–24.

Ó Tuathail, G. 2010. Localizing Geopolitics: Disaggregating Violence and Return in Conflict Regions. *Political Geography* 29 (5):256–65.

Ó Tuathail, G., and J. Agnew. 1992. Geopolitics and Discourse: Practical Geopolitical Reasoning in American Foreign Policy. *Political Geography* 11 (2):190–204.

Ó Tuathail, G., and S. Dalby. 1998. Introduction. In *Rethinking Geopolitics*, eds. Ó Tuathail, Gearóid and Simon Dalby, 1–15. New York: Routledge.

Ochieng, W. 2008. Kenyan Leaders Have Always Ignored the People's Voices. *The Daily Nation*, 18/02/2008.

Omondi, G. 2013. Uhuru's Diplomatic Snub Puts Kenya's Trade Interests at Risk. *The Daily Nation*, 25 October 2013.

Onyango-Obbo, C. 2012. Torn Between the East African Community and the Horn: The Tale of a 'Two-axis' Nation *The East African*, 02/06/2012.

Otiso, K. M., B. Derudder, D. Bassens, L. Devriendt, and F. Witlox. 2011. Airline Connectivity as a Measure of the Globalization of African cities. *Applied Geography* 31 (2):609–20.

Oyugi, W., P. Wanyande, and C. Odhiambo-Mbai eds. 2003. *The Politics of Transition in Kenya: From KANU to NARC*. Nairobi: Heinrich-Boell-Foundation.

Patton, M. Q. 1980. *Qualitative Evaluation Methods*. 2. print. ed. Thousand Oaks, Calif.: Sage.

Peck, J. 2011. Geographies of Policy: From Transfer-diffusion to Mobility-mutation. *Progress in Human Geography* 35 (6):773–97.

Peck, J., and N. Theodore. 2010. Mobilizing Policy: Models, Methods, and Mutations. *Geoforum* 41 (2):169–74.

Peck, J., and A. Tickell. 2002. Neoliberalizing Space. *Antipode* 34 (3):380–404.

Petrov, R., and P. Kalinichenko. 2011. The Europeanization of Third Country Judiciaries Through the Application of the EU Acquis: The Cases of Russia and Ukraine. *International & Comparative Law Quarterly* 60 (2):325–53.

Pierce, J., D. G. Martin, and J. T. Murphy. 2011. Relational Place-making: The Networked Politics of Place. *Transactions of the Institute of British Geographers* 36 (1):54–70.

Pirie, G. 2014. Geographies of Air Transport in Africa: Aviations 'Last Frontier'. In *The Geographies of Air Transport*, eds. Goetz, Andrew and Lucy Budd, 247–66. Farnham: Ashgate.

Power, M., and D. Campbell. 2010. The State of Critical Geopolitics. *Political Geography* 29 (5):243–46.

Power, M., G. Mohan, and M. Tan-Mullins. 2012. *China's Resource Diplomacy in Africa: Powering Development?* Basingstoke: Palgrave Macmillan.

Raffestin, C. 1977. Paysage et territorialité. *Cahiers de géographie du Québec* 21 (53–4):123–34.

Raffestin, C. 2012. Space, Territory, and Territoriality. *Environment and Planning D: Society and Space* 30 (1):121–41.

Richard, Y., and G. V. Hamme. 2013. L'Union européenne, un acteur des relations internationales. Étude géographique de l'actorness européenne. *L'Espace géographique* 42 (1):15–31.

Risse, T. 2012. Identity Matters: Exploring the Ambivalence of EU Foreign Policy. *Global Policy* 3:87–95.

Rogers, J. 2009. From 'Civilian Power' to 'Global Power': Explicating the European Union's 'Grand Strategy' Through the Articulation of Discourse Theory. *Journal of Common Market Studies* 47 (4):831–62.

Rose, G. 1997. Situating Knowledges: Positionality, Reflexivities and Other Tactics. *Progress in Human Geography* 21 (3):305–20.

Rovnyi, I., and V. Bachmann. 2012. Reflexive Geographies of Europeanization. *Geography Compass* 6 (5):260–74.

Rutazibwa, O. U. 2010. The Problematics of the EU's Ethical (Self)Image in Africa: The EU as an 'Ethical Intervener' and the 2007 Joint Africa-EU Strategy. *Journal of Contemporary European Studies* 18 (2):209–28.

Said, E. W. 1978. *Orientalism*. London: Penguin.

Schimmelfennig, F., and U. Sedelmeier. 2005. *The Europeanization of Central and Eastern Europe*. New York, NY: Cornell Univ. Press.
Schimmelfennig, F., and D. C. Thomas. 2008. Normative Institutionalism and EU Foreign Policy in Comparative Perspective. *UCD Dublin European Institute Working Paper* 08–12.
Scholvin, S., and P. Draper. 2012. The Gateway to Africa? Geography and South Africa's Role as an Economic Hinge Joint Between Africa and the World. *South African Journal of International Affairs* 19 (3):381–400.
Scott, J. W. 2011. Reflections on EU Geopolitics: Consolidation, Neighbourhood and Civil Society in the Reordering of European Space. *Geopolitics* 16 (1):146–75.
Sidaway, J. D. 2006. On the nature of the beast: re-charting political geographies of the European Union. *Geografiska Annaler* 88 B (1):1–14.
Smith, K. E. 2008. *European Union Foreign Policy in a Changing World.* 2nd ed. Cambridge: Polity.
Smith, M., S. Keukeleire, and S. Vanhoonacker. 2014. *The Diplomatic System of the European Union: Evolution, Change and Challenges* London: Routledge.
Smith, N. 1984. *Uneven Development: Nature, Capital and the Production of Space*. Oxford: Blackwell.
Swyngedouw, E. 1997. Neither Global nor Local: 'Glocalization' and the Politics of Scale. In *Spaces of Globalizatio : Reasserting the Power of the Local*, ed. Cox, Kevin R., 138–66. New York, NY: Guilford Press.
Tannous, I. 2013. The Programming of EU's External Assistance and Development Aid and the Fragile Balance of Power between EEAS and DG DEVCO. *European Foreign Affairs Review* 18 (3):329–54.
Taylor, L. 2013. The Case as Space: Implications of Relational Thinking for Methodology and Method. *Qualitative Inquiry* 19 (10):807–17.
Taylor, P. J. 2004. The New Geography of Global Civil Society: NGOs in the World City Network. *Globalizations* 1 (2):265–77.
Taylor, P. J. 2009. *International Organizations: Network Connectivities*. Globalization and World Cities Research Network, [cited 15/01/2015]. Available from http://www.lboro.ac.uk/gawc/datasets/da18.html.
Telò, M., and F. Ponjaert. 2013. *The EU's Foreign Policy: What Kind of Power and Diplomatic Action?* Farnham: Ashgate.
Thiemeyer, G. 2010. *Europäische Integration: Motive, Prozesse, Strukturen*. Köln: Böhlau.
Thrift, N. 2000. It's the Little Things. In *Geopolitical Traditions: A Century of Geopolitical Thought*, eds. Dodds, Klaus and David Atkinson, 380–87. London; New York: Routledge.
Toje, A. 2011. The European Union as a Small Power. *Journal of Common Market Studies* 49 (1):43–60.
Tonra, B. 2000. Denmark and Ireland. In *The Foreign Policies of European Union Member States*, eds. Manners, Ian and Richard Whitman, 224–42. Manchester: Manchester University Press.

Twitchett, K. J. 1976. *Europe and the World: The External Relations of the Common Market*. London: Europa for the David Davies Memorial Institute of International Studies.

Van Hüllen, V. 2012. Europeanisation through Cooperation? EU Democracy Promotion in Morocco and Tunisia. *West European Politics* 35 (1):117–34.

Vanhoonacker, S., and K. Pomorska. 2013. The European External Action Service and Agenda-setting in European Foreign Policy. *Journal of European Public Policy* 20 (9):1316–31.

Vidich, A. J., and G. Shapiro. 1955. A Comparison of Participant Observation and Survey Data. *American Sociological Review* 20 (1):28–33.

Vines, A. 2010. Rhetoric from Brussels and Reality on the Ground: The EU and Security in Africa. *International Affairs* 86 (5):1091–1108.

Vink, M. P., and P. Graziano. 2008. Challenges of a New Research Agenda. In *Europeanization: New Research Agendas*, eds. Graziano, Paolo and Maarten Peter Vink, 3–22. Basingstoke: Palgrave Macmillan.

Vollaard, H. 2014. Explaining European Disintegration. *JCMS: Journal of Common Market Studies* 52 (5): 1142–59.

Webb, E. J., D. T. Campbell, R. D. Schwartz, and L. Sechrest. 1966. *Unobtrusive Measures: Nonreactive Research in the Social Sciences*. Chicago: Rand McNally.

Wessels, W. 1982. European Political Cooperation: A New Approach to European Foreign Policy. In *European Political Cooperation: Towards a Foreign Policy for Western Europe*, eds. Allen, David, Reinhardt Rummel and Wolfgang Wessels, 1–20. London: Butterworth.

Wong, R. 2008. Foreign Policy. In *Europeanization : New Research Agendas*, eds. Graziano, Paolo and Maarten Peter Vink, 321–34. Basingstoke: Palgrave Macmillan.

Wong, R. 2011. The Europeanization of Foreign Policy. In *International Relations and the European Union*, eds. Hill, Christopher and Michael Smith, 149–70. Oxford: Oxford Univ. Press.

Wong, R. 2012. Model Power or Reference Point? The EU and the ASEAN Charter. *Cambridge Review of International Affairs* 25 (4):669–82.

Wong, R., and C. Hill. 2012. *National and European Foreign Policies: Towards Europeanization*. London: Routledge.

Yeung, H. W.-c. 2003. Practicing New Economic Geographies: A Methodological Examination. *Annals of the Association of American Geographers* 93 (2):442–62.

Yeung, H. W.-c. 2005. Rethinking Relational Economic Geography. *Transactions of the Institute of British Geographers* 30 (1):37–51.

Yeung, H. W.-c. 2009. Regional Development and the Competitive Dynamics of Global Production Networks: An East Asian Perspective. *Regional Studies* 43 (3):325–51.

Zielonka, J. 2006. *Europe as Empire: The Nature of the Enlarged European Union*. Oxford ; New York: Oxford University Press.

Zielonka, J. 2008. Europe as a Global Actor: Empire by Example? *International Affairs* 84 (3):471–84.

Index

References to figures are shown in *italics*. References to tables are shown in **bold**. References to footnotes consist of the page number followed by the letter 'n' followed by the number of the note, e.g. 9n2 refers to note no. 2 on page 9.

For Product Safety Concerns and Information please contact our EU
representative GPSR@taylorandfrancis.com
Taylor & Francis Verlag GmbH, Kaufingerstraße 24, 80331 München, Germany

www.ingramcontent.com/pod-product-compliance
Ingram Content Group UK Ltd.
Pitfield, Milton Keynes, MK11 3LW, UK
UKHW022131190926
475702UK00013B/1739
9781138546677